MODERN DEVICES THERMODYNAMICS: BATTERIES, FUEL CELLS AND SUPERCAPACITORS

Dritan Hoxha

ARCLER
PRESS

www.arclerpress.com

Modern devices Thermodynamics: Batteries, Fuel Cells and Supercapacitors

Dritan Hoxha

Arcler Press

2010 Winston Park Drive,

2nd Floor

Oakville, ON L6H 5R7

Canada

www.arclerpress.com

Tel: 001-289-291-7705
001-905-616-2116
Fax: 001-289-291-7601
Email: orders@arclereducation.com

ISBN: 978-1-77361-290-4 (Hardcover)

ABOUT THE AUTHOR

Dritan Hoxha has a Masters of Education degree in Physics and a four years integrated university diploma in Physics from University of Shkoder "Luigj Gurakuqi," Shkoder, Albania. He has 10 years of experience of teaching Physics. Currently, he is working as a teacher at Gulistan Foundation, Shkoder and Visiting Lecturer at University of Shkoder "Luigj Gurakuqi," Albania.

TABLE OF CONTENTS

List of Figures .... *xi*

List of Tables .... *xiii*

List of Abbreviations .... *xv*

Preface .... *xvii*

PART I: BATTERIES

Chapter 1 Introduction to Batteries 3

1.1. Overview 4

1.2. Operation and Working of Lithium Ion Battery 6

Chapter 2 Types of Batteries 9

2.1. Primary Batteries 10

2.2. Secondary Batteries 18

Chapter 3 Thermodynamics of Lithium Ion Batteries 25

3.1. Overview 26

3.2. Accelerating Rate Calorimeter (ARC) 28

3.3. Rate Of Generation of Heat 30

Chapter 4 Lithium Battery Structure 35

4.1. Construction of Battery 37

4.2. Electrodes Description 38

4.3. Anode 38

4.4. Cathode 39

4.5. Need Of Maintenance 43

Chapter 5 Thermodynamics Involved in Transport Process of Lithium Ion Batteries 45

5.1. Introduction 46

5.2. Transport of Heat 48
5.3. Conditions of Interface 50
5.4. Set of Complete Equations 53

Chapter 6 Thermal Heat and Entropy related to Lithium Batteries 57
6.1. Thermal Heat Introduction 58
6.2. Entropy Introduction 60
6.3. Generation of Heat 62
6.4. Calculation of Heat 64
6.5. Thermal Runway of Battery For Its Safety 66

Chapter 7 Applications of Batteries in Modern World 69
7.1. Aircraft 72

PART II FUEL CELLS

Chapter 8 Overview 77
8.1. Introduction 78
8.2. Emergence 80
8.3. Problems In Emergence of Fuel Cell 83
8.4. Difference Between Fuel Cell And Battery 84

Chapter 9 Electrochemical Aspects of Fuel Cells 89
9.1. Energy And Heat Balance System 90
9.2. Change In Gibb's Free Energy 92
9.3. Principles Of Thermodynamics Related To Fuel Cells 95
9.4. Fuel Cell Reactions 98

Chapter 10 Working and Operation of Fuel Cells 99
10.1. Reversible Cell Potential 100
10.2. Effect Of Temperature On Reversible Cell Potential 105
10.3. Effect Of Pressure On Reversible Cell Potential 108

Chapter 11 Efficiencies and Energy Losses 111
11.1. Efficiency Of Fuel Cells 112
11.2. Efficiency Of Carnot Cycle 113
11.3. Comparison Of The Efficiencies Of Fuel Cell And Carnot Cycle 115

11.4. Losses Of Fuel Cells 117
11.5. Reversible Energy Loss Compared To Energy Efficiency 118
11.6. Irreversible Energy Loss Compared To Energy Efficiency 119

Chapter 12 Types of Fuel Cells 123
12.1. Solid Oxide Fuel Cell (SOFC) 124
12.2. Phosphoric Acid Fuel Cell (PAFC) 129
12.3. Polymer Electrolyte Membrane Fuel Cell (PEMFC) 135
12.4. Alkaline Membrane Fuel Cell (AFC) 137
12.5. Molten Carbonate Fuel Cell (MCFC) 144

Chapter 13 Applications of Fuel Cell 151
13.1. Power 152
13.2. Cogeneration 152
13.3. Automobiles 153
13.4. Buses 154
13.5. Forklifts 154
13.6. Motorcycles And Bicycles 154
13.7. Airplanes 155
13.8. Submarines 155
13.9. Portable Power Systems 155

PART III SUPER CAPACITORS

Chapter 14 Introduction to Super Capacitors 159
14.1. Electrolytic Capacitors Before Super Capacitors 160
14.2. Emergence/Background 163
14.3. Overview 164
14.4. Construction And Working 165
14.5. Energy And Power 170
14.6. Devices And Technology Of Super Capacitors 170
14.7. Winding Technology 171

Chapter 15 Properties of Super Capacitors 173
15.1. Equivalent Circuit Of Super Capacitors 174
15.2. Plot Of Ragone 176
15.3. Power Availability 178

15.4. Energy Availability 179
15.5. Double Layer Capacitors (DLC) 180
15.6. Performance 192
15.7. Distinction Between Super Capacitor And Capacitor 194

Chapter 16 Taxonomy and Characteristics of Super Capacitors 199
16.1. Pseudo Capacitors 201
16.2. Hybrid Capacitors 203
16.3. High Capacitance Electrodes 205
16.4. Quantitatively Modeling 208
16.5. Measurement Of Capacitance Of Super Capacitors 210

Chapter 17 Super Capacitors as Alternative Energy Source and Its Applications 213
17.1. Overview 214
17.2. Materials 214
17.3. Discharge Cycle 216
17.4. Advantages 217
17.5. Disadvantages 218
17.6. Applications 219
17.7. Advantages When Super Capacitor Combined With Battery 223
17.8. Advancement In Superconductors 225

References 227

Index 245

LIST OF FIGURES

Figure 2.1. Simple alkaline manganese structure.

Figure 2.2. Button batteries.

Figure 3.1. Schematic diagram of heat flux sensor.

Figure 4.1. Drawing of lead acid battery.

Figure 6.1. Graphs for OCV as a function of temperature for battery at different SOC (a) SOC = 0.122 (b) SOC = 0.458 (c) SOC = 0.644 (d) SOC = 0.813.

Figure 9.1. Ideal versus actual performance of fuel cell.

Figure 10.1. Thermodynamic model of fuel cell.

Figure 10.2. Effect of temperature on fuel cell.

Figure 11.1. Efficiency of Carnot cycle.

Figure 11.2. Polarization curve.

Figure 12.1. Hydrogen oxygen fuel cell.

Figure 12.2. Working of MCFC.

Figure 14.1. Simple circuit.

Figure 15.1. Energy density vs. power density.

Figure 15.2. Interface of electrode and electrolyte.

Figure 15.3. Simple construction of EDLC.

Figure 15.4. Leakage current RC circuit.

Figure 16.1. Energy vs. power plot for energy storage devices.

Figure 16.2. Taxonomy of super capacitors.

Figure 16.3. Schematic of surface oxygen containing groups.

Figure 16.4. Relation between evolved C and CO.

Figure 17.1. Discharge cycle of super capacitor.

Figure 17.2. Schematic diagram of EDLC.

LIST OF TABLES

Table 15.1. Performance Evaluation of G and RMGO in Super Capacitor.

LIST OF ABBREVIATIONS

$\wp_{(s,FC)}$	Entropy production in fuel cell
$\wp_{(s,HE)}$	Entropy production in heat engine
$\wp_s$	Entropy production (J/mol)
$\wp_{s,TER}$	Entropy production in thermal energy reservoir
Δ	A finite difference of a property
μ	Chemical potential
0	Standard temperature and pressure condition
Act	Activation overpotential
Aux	Auxiliary
C.V.	Control volume
conc.	Concentration
c_p	Average specific heat at constant pressure (J/molK)
CP	Specific heat at constant pressure
E	Cell potential (V)
E	Electric charge per electron (=1.6023 × 10^{-19} Coulomb/electron)
elec.	Electrical Work
E_r	Reversible cell potential (V)
E_r^0	Standard reversible cell potential (V)
EX	Exhaust
F	Faraday constant (=96,487 Coulomb/mole electron)
F	Fuel
FC	Fuel Cell
G	Gas
G	Gibbs free energy (J)
G	Specific Gibbs free energy (J/mol)
H	Enthalapy (J)
H	High temperature
H	Efficiency

$h_{(f,i)}$	Specific enthalpy of formations of species i (J/mol)
$h_{(f,i)}^{o}$	Specific enthalpy of formations of species i at reference conditions (J/mol)
$h_{(s,i)}$	Specific sensible of enthalpy of species i (J/mol)
HHV	Higher heating value (J)
h_p,h_R	Specific enthalpy of products and reactants (J/mol)
I	Current
I	The chemical species
I_0	Exchange current (A)
J_o	Exchange current density (A/m^2)
K	Equilibrium constant
l	Liquid
L	Low Temperature
LHV	Lower heating value (J)
Max	Maximum
Mi	Chemical formula for species i
N	Number of mole electron
N	Number of moles
$N^{\cdot}$	Rate of mole (mole/s)
Ohm	Ohmic resistance
OX	Oxidation
P	Pressure (pas)
P	Products of combustion
Q	Heat (J/mole)
$Q^{\cdot}$	Rate of heat (W)
R	Difference between products and reactants
R	Universal gas constant
S	Entropy
S	Entropy (J/molK), stoichiometry
T	Temperature (K)
tn	Thermoneutral overpotential
V	Stoichiometric reaction coefficient
v_F’	Number of moles of fuel
v_i’, v_i”	Number of moles for species i in the reactant and products mixture
W	Specific work (J/mol)
$W^{\cdot}$	Power (W)
X	Mole fraction

PREFACE

Presently our planet faces gigantic vitality challenges from the shortage of petroleum derivatives and the synchronous nursery impacts. Consuming of petroleum derivatives with substantial anthropogenic CO_2 emanations outpaces nature's reusing capacity, bringing about noteworthy ecological damage, for example, a dangerous atmospheric deviation and seas fermentation. Therefore, vitality frailty, rising costs of petroleum products and atmosphere changes risk surprisingly our wellbeing and political solidness. Thus, we are watching an expansion in sustainable power source advancements from sun based, twist and in addition the approach of half and half electric vehicles with low outflows. In any case, because of the irregular character of sustainable, dependable electrical vitality stockpiling frameworks are required for adjusting these advancements to the request in power. Among the different sensible arrangements, vitality can specifically be put away electrochemically in Batteries, Fuel Cells and Super capacitors. Despite the fact that batteries as of now show substantially higher vitality thickness, their generally low power thickness and poor cycle life thwart high power requesting applications, for example, regenerative braking and load leveling frameworks.

In the previous decade, there have been energizing advancements in the field of lithium particle batteries as vitality stockpiling gadgets, bringing about the utilization of lithium particle batteries in zones extending from little convenient electric gadgets to substantial power frameworks, for example, half and half electric vehicles. Be that as it may, the greatest vitality thickness of current lithium particle batteries having topotactic science isn't sufficient to meet the requests of new markets in such territories as electric vehicles. Along these lines, new electrochemical frameworks with higher vitality densities are being looked for, and metal-air batteries with change science are viewed as a promising hopeful. All the more as of late, encouraging electrochemical execution has driven much research enthusiasm for Li-air and Zn-air batteries. This audit gives a review of the basics and late advance in the region of Li-air and Zn-air batteries, with the point of giving a superior comprehension of the new electrochemical frameworks.

The battery is a fabulous however significantly misconstrued wellspring of compact power. It is an electrical power aggregator that stores a particular measure of electrical vitality, the sum being subject to its electrical size, or limit. As a water weight tank can supply a specific measure of gallons every moment for a timeframe relying upon its size, so a battery may convey a specific measure of amperes for a timeframe relying upon its size. This is known as the ampere–hour (A.H.) limit of a battery. Because of the high vitality/control thickness, in connection to the weight and volume of Lithium-particle (Li^+) battery innovation, there are some waiting security concerns while charging and releasing the batteries. Despite the fact that an officially develop innovation, upgrades to Li+ battery operation are progressing. This application note depicts some of those enhancements. It additionally shows different charging control plans to guarantee that the cells are appropriately charged utilizing steady present, consistent voltage (CCCV) approaches. A few charging circuits show approaches for single-cell and various cell L^+ chargers.

Warmth motors in view of non-renewable energy source burning produce destructive contaminations and ozone depleting substance emanations. Natural concerns and practical advancement call for new innovation for vitality change and power age, which is more effective, earth well-disposed and good with elective powers and sustainable power sources and bearers. Power devices meet every one of these necessities, and are being produced as one of the essential vitality advances without bounds. In this section, the thermodynamic execution of power modules is brake down, vitality transformation productivity of energy units and warmth motors is contemplated and looked at, and misguided judgments about power module proficiency illuminated. It is demonstrated that both energy components and warmth motors have a similar most extreme hypothetical effectiveness, which is comparable to the Carnot proficiency, while working on a similar fuel and oxidant. In any case, energy units are free from the high temperature confine forced by materials on warm motors and less irreversibility's related with warm dismissal. As a result, power modules can have higher handy efficiencies.

By differentiate, Super capacitors store bigger measures of vitality than the customary dielectric capacitors and give vitality far speedier than batteries. Accordingly, they are especially adjusted for applications requiring vitality beats in brief timeframes, e.g., seconds or several seconds. Such excellent properties emerge from the nanometric scale capacitors shaped by the captivated terminal material and a layer of pulled in particles from the electrolyte on its surface. The thickness of the anode electrolyte interface is specifically controlled by the extent of particles. As of late, Super capacitors have been proposed and generally advertised for different applications. Coupled for instance with a battery/inside burning motor in crossover vehicles, Super capacitors enhance the battery lifetime/mileage and the vitality recuperation effectiveness in braking. They can likewise balance out current when discontinuous sustainable power sources are presented in the lively blend. Despite the fact that Super capacitors are currently industrially accessible, regardless they require upgrades, particularly to enhance their vitality thickness and cut the cost in the meantime. It requires a major comprehension of their properties and correct working standards, notwithstanding enhancing cathode materials, electrolytes and coordination in frameworks. Every one of these subjects prompted an extremely solid inspiration of scholastics and industry amid the most recent decade.

The vitality thickness of Super capacitors can be upgraded by expanding the voltage or potentially the capacitance. For achieving these goals, different methodologies have been proposed in the writing, including the advancement of new materials, new geometries and new electrolytes. Be that as it may, the execution and the cost of the super capacitor gadget are the primary parameters which direct the decisions of industry. Therefore, a large portion of the Super capacitors by and by accessible available depend on enacted carbon (AC) anodes and a natural electrolyte. The primary points of interest of enacted carbon are identified with its high flexibility of structure/surface, minimal effort and very created surface region. More critical, its high electrical conductivity permits the acknowledgment of high power frameworks without requiring planning convoluted, and thus costly, anode materials. Albeit for the most part antagonistic, because of the utilization of acetonitrile as dissolvable, the natural electrolytes are favored by industry to the watery ones as a result of their high steadiness window, e.g., 2.7–2.8 V, enabling high vitality thickness to be come to. Despite the fact that the symmetric AC/AC innovation in natural electrolyte is presently develop, it can be as yet enhanced by growing new permeable carbons as well as electrolyte details. Next to, elective arrangements in view of different sorts of electrolytes must be explored keeping in mind the end goal to grow either all the more performing frameworks or less expensive ones.

PART I: BATTERIES

CHAPTER 1

INTRODUCTION TO BATTERIES

CONTENTS

1.1. Overview.. 4
1.2. Operation and Working of Lithium Ion Battery 6

1.1. OVERVIEW

A lot of petroleum derivative is devoured each day to create power and power ignition motors, causing worldwide worry about vitality efficiency, ozone depleting substance outflow, and consumption of regular assets. Subsequently, generous effort has been made to create sustainable power source innovation, for example, sun based boards, energy units, and electrically fueled vehicles. These all require techniques for electrical vitality stockpiling. Batteries, as gadgets for putting away vitality artificially, have points of interest of high transportability, high change efficiency, long life, and zero fumes discharge. They are perfect power hotspots for versatile gadgets, autos, and reinforcement control supplies. Consequently, creating battery innovation, especially rechargeable batteries, has turned into a key issue for science and industry.

The battery is a fundamental part of all flying machine electrical frameworks. Batteries are utilized to begin motors and assistant power units, to give crisis reinforcement energy to basic flight gear, to guarantee no-break control for route units and fly-by-wire PCs, and to give ground control capacity to upkeep and preflight checkouts. Huge numbers of these capacities are mission basic, so the execution and unwavering quality of an air ship battery is of impressive significance. Other vital prerequisites incorporate ecological roughness, a wide working temperature go, simplicity of support, fast revive ability, and resistance to mishandle [1].

Truly, just a couple of sorts of batteries have been observed to be appropriate for air ship applications. Until the 1950s, vented lead-corrosive (VLA) batteries were utilized only. In the late 1950s, military airplane started changing over to vented nickel-cadmium (VNC) batteries, principally due to their unrivaled execution at low temperature. The VNC battery in this manner discovered far-reaching use in both military and business flying machine. The main other kind of battery utilized amid this period was the vented silver-zinc battery, which gave a vitality thickness around three times higher than VLA and VNC batteries. This battery sort was connected to a few sorts of U.S. Aviation based armed forces fighters (F-84, F 105, and F-106) and U.S. Naval force helicopters (H-2, H-13, and H-43) in the 1960s. Albeit silver-zinc air ship batteries were alluring for diminishing weight and size, their utilization has been ceased because of poor unwavering quality and high cost of possession.

In the late 1960s and mid 1970s, a broad advancement program was directed by the U.S. Aviation based armed forces and Gulton Industries to qualify fixed nickel-cadmium (SNC) flying machine batteries for military and business applications. This battery innovation was effectively exhibited on a Boeing KC-135, a Boeing 727, and an UH-1F helicopter. Prior to the innovation could be progressed into generation, be that as it may, Gulton Industries was assumed control by SAFT and a choice was made to end the program.

In the late 1970s and mid-1980s, the U.S. Naval force spearheaded the improvement of fixed lead-corrosive (SLA) batteries for air ship applications. SLA batteries were at first connected to the AV-8B and F/A-18, bringing about a significant dependability and practicality (R&M) change contrasted and VLA and VNC batteries. The Navy in this way changed over the C-130, H-46, and P-3 to SLA batteries. The U.S. Flying corps took after the Navy's lead, changing over various flying machine to SLA batteries, including the A-7, B-1B, C-130, C-141, KC-135, F-4, and F-117. The expression "High Reliability, Maintenance-Free Battery," or HRMFB, was begat to accentuate the enhanced R&M ability of fixed cell flying machine batteries. The utilization of HRMFBs soon spun off into the business part, and various business and general avionics air ship today have been retrofitted with SLA batteries.

In the mid-1980s, impelled by expanding requests for HRMFB innovation, a restored enthusiasm for SNC batteries occurred. A program to create progressed SNC batteries was started by the U.S. Aviation based armed forces, and Eagle-Picher Industries was contracted for this exertion [Flake, 1988; Johnson et al., 1994]. The B-52 plane was the first air ship to retrofit this innovation. SNC batteries likewise have been created by ACME for a few airplane applications, including the F-16 fighter, Apache AH-64 helicopter, MD-90, and Boeing 777 [2].

A current advancement in airplane batteries is the "low upkeep" or "ultra-low support" nickel cadmium battery. This battery is planned to be an immediate substitution of customary VNC batteries, dodging the need to supplant or change the charging framework. In spite of the fact that the battery still requires planned support for electrolyte filling, the upkeep recurrence can be diminished significantly. This sort of battery has been a work in progress by SAFT and all the more as of late by Marathon. Constrained flight tests have been performed by the U.S. Naval force on the H-1 helicopter. Utilization of this innovation to business airplane is likewise being sought after.

Deciding the most appropriate battery sort and size for a given air ship sort requires nitty gritty information of the application prerequisites, stack profile, obligation cycle, natural components, and physical limitations and the attributes of accessible batteries, execution capacities, charging necessities, future, and cost of proprietorship. With the different battery sorts accessible today, significant aptitude is required to measure, select, and get ready specialized specifications.

1.1.1. Definition

The battery is a phenomenal however enormously misconstrued wellspring of versatile power. It is an electrical power gatherer that stores a particular measure of electrical vitality, the sum being subject to its electrical size, or limit. As a water weight tank can supply a specific measure of gallons every moment for a timeframe relying upon its size, so a battery may convey a specific measure of amperes for a timeframe relying upon its size. This is known as the ampere–hour (A.H.) limit of a battery.

A battery is an extremely proficient gadget. On the off chance that 10 A.H's. are utilized, 10 A.H. come back to the battery by reviving will take the battery back to a condition of full charge. This is much the same as a weight tank where 10 ladies per minutes was utilized for 10 minutes. One hundred gallons would have been utilized, so 100 gallons would refill the tank again.

1.2. OPERATION AND WORKING OF LITHIUM ION BATTERY

A concoction Redox response includes the exchange of electrons starting with one animal categories then onto the next (generally metals). Without the battery setup, the response would simply continue and create no electrical work. By running the electrons through an outer circuit the response can create electrical, i.e., non-PdV work.

$$Zn \rightarrow Zn^{+2} + 2e^-$$

$$Cu^{+2} + 2e^- \rightarrow Cu$$

Net reaction

$$Zn(s) + Cu^{+2}(aq) \rightarrow Zn^{+2}(aq) + Cu(s)$$

1.2.1. Gibbs Energy and K Relationship

The Gibbs stage control of thermodynamics can be connected to battery cells, gave the framework is under thermodynamic harmony. The quantity of degrees of opportunity f is given by

$$f = c - p + 2 \qquad (1)$$

where c is the quantity of segments and p is the quantity of existing together stages. Regularly, the lithium intercalation in the anode is a two-segment framework (c = 2), the Li^{+} particles and the host structure. In this way, if giving a solitary stage (p = 1), other than temperature (T) and weight (P), there is just a single more level of flexibility (f = 3). In this way, the substance potential μ (or OCV) must be an element of T, P, and organization x. On account of a two-stage blend, there is not any more autonomous level of opportunity other than T and P (f = 2), so all thermodynamic capacities, for example, μ, S, or H ought to stay steady with creation x under a fixed T and P [3].

This net response composed above has a K and a Gibbs Energy

$$\Delta G^0_{rxn} = \Delta G^0_f\left(Zn^{+2}(aq)\right) - \Delta G^0_f\left(Cu^{+2}(aq)\right) = -147.1 - 65.5 = -212.6\frac{kJ}{rxn} \qquad (2)$$

$$\ln K = -\frac{\Delta G_0}{RT_0} = \frac{197}{2.48} = 85.5 \qquad (3)$$

Electrons move from a district of moderately negative potential to one of more positive potential since they moved from the Zn (metal) to the Cu particle. The adjustment in potential must be with the end goal that the adjustment in vitality of the electrons is indistinguishable to the adjustment in vitality of the framework. It is only an alternate point of view on a similar response [4].

The potential (electrical) vitality for a charge, q, in a potential is

$$E = q\Phi \, or \, \Delta E = q\Delta\Phi \qquad (4)$$

On the off chance that we know the charge on an electron, which is negative, then

$$q = -\boldsymbol{nF} \qquad (5)$$

F is the Faraday, which, in SI units, is the quantity of coulombs per mole of electrons, and n is the quantity of moles of electrons where F = 96,500 Coul/mole. The potential is in volts, and the charge is in coulombs and the vitality is in Joules. A Volt is one Joule for every Coulomb [5].

CHAPTER 2

TYPES OF BATTERIES

CONTENTS

2.1. Primary Batteries 10

2.2. Secondary Batteries 18

2.1. PRIMARY BATTERIES

In spite of the fact that various anode-cathode mixes can be utilized as essential battery frameworks, just a moderately few have made viable progress. Zinc has been by a wide margin the most well-known anode material for essential batteries due to its great electrochemical conduct, high electrochemical proportionality, similarity with fluid electrolytes, sensibly great timeframe of realistic usability, minimal effort, and accessibility. Aluminum is alluring a result of its high electrochemical potential and electrochemical comparability and accessibility, however because of passivation and for the most part constrained electrochemical execution, it has not been produced effectively into a down to earth dynamic essential battery framework. It is presently being considered in mechanically rechargeable or refuelable aluminum/air batteries and for possible later use battery frameworks. Magnesium additionally has appealing electrical properties and ease and has been utilized effectively in a dynamic essential battery, especially for military applications, on account of its high vitality thickness and great time span of usability. Business intrigue has been restricted. Magnesium additionally is mainstream as the anode for possible later use batteries. Presently there is an expanding center around lithium, which has the most astounding gravimetric vitality thickness and standard capability of the considerable number of metals. The lithium anode battery frameworks, utilizing various distinctive non-aqueous electrolytes in which lithium is steady and diverse cathode materials, offer the open door for higher vitality thickness and different advances in the execution attributes of essential frameworks [6].

2.1.1. The Early Batteries

The essential battery is an advantageous wellspring of energy for versatile electric and electronic gadgets, lighting, photographic hardware, PDA's (Personal Digital Assistant), correspondence gear, listening devices, watches, toys, memory reinforcement, and a wide assortment of different applications, giving opportunity from utility power. Real favorable circumstances of the essential battery are that it is helpful, basic, and simple to utilize, requires pretty much nothing, assuming any, support, and can be measured and formed to fit the application. Other general focal points are great time span of usability, sensible vitality and power thickness, dependability, and satisfactory cost. Essential batteries have existed for more than 100 years, yet up to 1940, the zinc-carbon battery was the just a single in wide utilize. Amid World War II and the after a long time frame, noteworthy advances were made, with the zinc-carbon framework, as well as with new and prevalent sorts of batteries. Limit was enhanced from under 50 Wh/kg with the early zinc-carbon batteries to more than 400 Wh/kg now acquired with lithium batteries. The timeframe of realistic usability of batteries at the season of World War II was constrained to around 1 year when put away at direct temperatures; the time span of usability of present-day regular batteries is from 2 to 5 years. The time span of usability of the more current lithium batteries is as high as 10 years, with a capacity of capacity at temperatures as high as 70°C. Low temperature operation has been stretched out from 0 to 40°C, and the power thickness has been enhanced many overlay. Uncommon low-deplete batteries utilizing a strong electrolyte have timeframes of realistic usability in overabundance of 20 years [7].

A significant number of the huge advances were made amid the 1970–90 time frame and were fortified by the simultaneous advancement of electronic innovation, the new requests for versatile power sources and the help for the space, military, and the earth change programs. Amid this period, the zinc/basic manganese dioxide battery started to supplant the zinc carbon or Leclanche battery as the main essential battery, catching the significant offer of the US advertise. Natural concerns prompted the disposal of mercury in many batteries with no hindrance of execution, yet additionally prompted the eliminating of those batteries, zinc/mercuric oxide and cadmium/mercuric oxide, that utilized mercury as the cathodic dynamic material. Luckily, zinc/air and lithium batteries were created that could effectively supplant these "mercury" batteries in numerous applications. A noteworthy achievement amid this period was the advancement and promoting of various lithium batteries, utilizing metallic lithium as the anode dynamic material. The high particular vitality of these lithium batteries, no less than twice that of most ordinary watery essential batteries, and their predominant timeframe of realistic usability opened up an extensive variety of uses—from little coin and barrel shaped batteries for memory reinforcement and cameras to expansive batteries which were utilized for move down power for rocket storehouses. Increments in the vitality thickness of essential batteries has decreased amid the previous decade as the current battery frameworks have developed and the improvement of new higher vitality batteries is restricted by the absence of new and additionally untried battery materials and sciences. In any case, progresses have been made in other critical execution qualities, for example, control thickness, time span of usability and wellbeing. Cases of these current advancements are the powerful zinc/antacid/manganese dioxide batteries for convenient shopper gadgets, the change of the zinc/air battery and the presentation of new lithium batteries. These enhanced attributes have opened up numerous new open doors for the utilization of essential batteries. The higher vitality thickness has brought about a significant diminishment in battery size and weight. This decrease, brought with the advances in gadgets innovation, has made numerous convenient radio, correspondence, and electronic gadgets down to earth. The higher power thickness has made it conceivable to utilize these batteries in PDA's, handsets, correspondence, and reconnaissance hardware, and other high-control applications, that to this point must be fueled by optional batteries or utility power, which don't have the comfort and flexibility from upkeep and energizing as do essential batteries. The long time span of usability that is presently normal for some essential batteries has also brought about new uses in restorative hardware, memory reinforcement, and other long haul applications and in a change in the lifetime and unwavering quality of battery worked gear. The overall essential battery advertise has now achieved more than $20 billion every year, with a development rate surpassing 10% every year. Most by far of these essential batteries are the natural barrel shaped and level or catch sorts with limits underneath 20 Ah. Few bigger essential batteries, extending in examine to a few thousand ampere-hours, are utilized as a part of flagging applications, for standby power and in other military and exceptional applications where autonomy from utility power is obligatory [8].

2.1.2. Alkaline Batteries

The soluble battery is an enhanced dry cell. The half-responses are comparative, yet the electrolyte is a fundamental KOH glue, which disposes of the development of gasses and keeps up the Zn terminal. Its applications are the same with respect to dry cell. Since its presentation in the mid-1960s, the antacid manganese dioxide (zinc/KOH/MnO_2) battery has turned into the predominant battery framework in the convenient battery advertise. This came to fruition in light of the fact that the antacid framework is perceived as having a few points of interest over its acidic electrolyte partner, the Leclanche' or zinc-carbon battery, the previous market pioneer with which it contends. The favorable circumstances and weaknesses of basic manganese dioxide batteries contrasted with zinc-carbon batteries are very difficult to achieve. The soluble manganese dioxide battery is accessible in two plan arrangements:

1. as generally expansive size barrel shaped batteries.
2. as smaller than expected catch batteries [9].

There are likewise different cell batteries produced using different sizes of unit cells. While the basic cell is as yet experiencing change, a few advancements in the development of the present barrel shaped cell innovation are especially remarkable. After the underlying ideas of a gelled/amalgamated zinc powder anode in a focal compartment and utilization of vented plastic seals had been set up, the principal real progress was the butt-crease metal complete which enabled the cell to have more noteworthy inside volume. Next came the disclosure that natural inhibitors could lessen the rate of gassing caused by contaminants in the zinc anode, bringing about an item with decreased lump and spillage. Another real improvement was the presentation of the plastic mark complete and bring down profile seal, which allowed a further huge increment in the inside volume accessible for dynamic material and a significant increment in the limit of the battery. Maybe the most noteworthy change to the basic cell started in the mid-1980s with the continuous decrease of the measure of mercury in the anode and the improvement of cells containing no additional mercury. This pattern, which was helped by a considerable change in the unwavering quality of cell materials coming about because of decreased pollution levels, was driven by overall worry over the ecological effect of the materials utilized as a part of batteries after their transfer. Improvements, for example, these have empowered the soluble MnO_2 battery to pick up as much as a 60% expansion in particular vitality yield since first experience with keep pace with the necessities of the purchaser. Its authority position should bolster encourage mechanical upgrades, which will guarantee proceeded with advertise predominance [10].

2.1.2.1. Construction

Figure 2.1 demonstrates the development of average round and hollow soluble manganese dioxide batteries from two makers. A round and hollow steel can be the holder for the cell. It additionally fills in as the cathode current gatherer. The cathode, a compacted blend of manganese dioxide, carbon, and conceivably different added substances, is situated inside the can as an empty barrel in close contact with the can internal surface. The cathode can be framed by straightforwardly forming it in the can. On the other hand, rings of cathode material can be shaped outside the phone and after that pushed into the can. Inside the empty focal point of the cathode are put layers of separator material. Within that is the anode, with a metal authority reaching it, and making association through a plastic seal to the negative terminal of the cell. The cell has best and base spreads and a metal or plastic coat connected. The spreads fill a double need. Other than giving a beautiful and erosion safe complete, they likewise accommodate the correct extremity of the battery. This is essential in light of the fact that the round and hollow soluble manganese battery is utilized as an immediate substitution for Leclanche' batteries. Leclanche' batteries have a level contact on the negative (zinc can) end, and a catch contact on the positive end to suit the carbon bar utilized as present gatherer. The barrel shaped antacid manganese dioxide cell is constructed "inside-out" in connection to the Leclanche' cell, with the cell holder as the positive ebb and flow authority and the finish of the negative gatherer jutting from the focal point of the seal. Along these lines to give it an outer frame like the Leclanche' battery, the round and hollow soluble battery must utilize a level cover to contact the end of the negative authority, and a base cover containing the Leclanche' positive projection in contact with the base of the can [11]. A few makers form the bulge into the can itself, and hence needn't bother with the base cover.

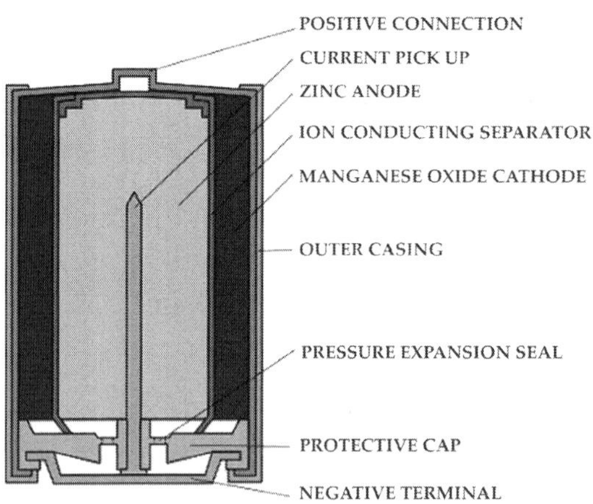

Figure 2.1: Simple alkaline manganese structure. (Source: *https://upload.wikimedia.org/wikipedia/commons/thumb/0/04/Alkaline-battery-english.svg/220px-Alkaline-battery-english.svg.png*).

2.1.3. Button Batteries

The catch cell, otherwise called coin cell, empowered conservative plan in versatile gadgets of the 1980s. Higher voltages were accomplished by stacking the cells into a tube. Cordless phones, restorative gadgets and security wands at airplane terminals utilized these batteries.

Albeit little and modest to construct, the stacked catch cell dropped out of support and offered approach to more traditional battery groups. A disadvantage of the catch cell is swelling if charged too quickly. Catch cells have no security vent and must be charged at a 10- to 16-hour charge; be that as it may, more up to date plans guarantee quick charge capacity.

Figure 2.2: Button batteries. (Source: *https://upload.wikimedia.org/wikipedia/commons/c/c7/Coin-cells.jpg*).

Most catch cells being used today are non-rechargeable and are found in therapeutic inserts, watches, listening devices, auto keys and memory reinforcement. Figure 2.2 shows the catch cells with a cross area.

2.1.4. Metal Air Batteries

Metal-air batteries are a sort of battery discernable by the way that one of the electroactive materials (oxygen) does not should be put away. This hypothetically streamlines the plan and expands the vitality thickness of the phone, and these batteries are additionally alluded to as power device/battery crossovers. This innovation can possibly accomplish the most elevated particular vitality of any known battery innovation. The main metal-air batteries were created in 1868, based around a MnO_2/carbon cathode, before more present-day plans were designed in 1932. From that point forward many metals have been discovered fitting for utilize, including Ca, Al, Fe, Cd, and Zn. The primary rechargeable Li-air battery was made in 1996, in spite of the fact that it was not until 2006 that noteworthy research started into this cell innovation.

In the course of the most recent decade, Zn-air, and Li-air have pulled in the most consideration and the figures in this reality sheet depend on current condition of threat with these advancements. Basically, metal-air batteries comprise of an uncovered permeable carbon terminal – the air cathode, which traps oxygen gas isolated from the metal anode by an electrolyte which might be strong, watery, non-aqueous (natural), or a crossover. Starting at 2011 the non-watery electrolyte sort is the farthest progressed [12].

The main thrust behind the huge research exertion as of now in progress is an endeavor to open the special level of vitality thickness that Li-air batteries guarantee. Zn-air batteries are substantially less expensive and ecologically more secure than Li-air in this way hold leeway in spite of having a lower utmost of particular vitality. Additionally, while low-limit metal-air batteries that are as of now available are generally minimal effort, electrical energizing is troublesome and wasteful; auxiliary (rechargeable) batteries being worked on at present have a lifetime of a couple of hundred cycles and commonly achieve just half cycle proficiency.

For Li-batteries, air drying out layers are required as Li is exceptionally receptive with dampness, and gathering of cell parts must be done under latent conditions. This is distinctive to the development of Zn based batteries, where the material is steady in surrounding conditions [13].

Zinc-air batteries are ecologically amiable. As indicated by a provider of Zn-air batteries for use in U.S. military equipment, there are no extraordinary care or taking care of prerequisites and the batteries are naturally sheltered being used, capacity, transportation, and transfer. Be that as it may, Lithium responds unequivocally with water, and Li-air batteries must be shielded from surrounding dampness amid operation and capacity, while keeping up a decent supply of oxygen amid operation.

2.1.5. Lithium Primary Batteries

Created in the mid-1970s, essential lithium batteries are the most vitality thick electrochemical cells made for watches, film cameras, therapeutic gadgets, and military purposes. Lithium essential cells have a run of the mill gravimetric thickness 250 Wh kg, against just 150 Wh kg for Li-particle batteries. Different innovations that vary in science and development have been utilized to create essential lithium batteries. They are classified into three gatherings, as indicated by the shape and the sort of cathode and electrolyte utilized. Ice and Sullivan said that in 2009, essential batteries including soluble that were driving the market, carbon-zinc, and lithium cells made 23.6 % of the worldwide market with 3 % for essential lithium cells; it has been anticipated a 7.4 % decrease in 2015 because of the improvement of rechargeable batteries. Concurring the Freedonia's examinations, the generation of essential batteries multiplied in the vicinity of 2002 and 2012. US interest for essential and optional batteries are relied upon to grow 4.2 % every year to $17.1 billion out of 2017. Lithium batteries will offer the best development openings in both the rechargeable and essential battery fragments. Figure 2.1 presents the Japanese battery showcase in the year 2013, in which lithium-metal innovation is 17 % portion of the aggregate battery income [14].

By putting an extremely unadulterated lithium-metal thwart as anode component and a lithium salt in a nonaqueous arrangement as electrolyte, another age of electrochemical generators was conceived in the mid-1960s. Essentially, the charge transport is indistinguishable to nickel metal hydride (Ni-MH) or nickel-cadmium (Ni-Cd) batteries, aside from that Li^+ particles are made by the straightforward reaction

$$Li \rightarrow Li^+ + e^- \qquad (1)$$

freeing one electron through the outer circuit and one particle brought into the permeable structure of the cathode. As talked about in the first section, the principle parameters of vitality stockpiling frameworks are vitality thickness (gravimetric and volumetric), control thickness, vitality efficiency, and vitality quality. The considerable appeal of the lithium innovation originates from the way that Li is a lighter metal molar weight Mw¼6.941 g moll and thickness 0.51 g cm^3, with the electronic configuration $(He)_2s_1$. The specific limit of Li metal is 3860 mAh g1 and the couple Li_0 has the most elevated electro activity with standard redox potential 3.04 V against H_2/ H^+. Thusly, the voltage of lithium batteries is likewise significantly higher than that of the Pb-corrosive and Ni-metal hydride, since lithium is the most electropositive component found in nature. Essential and auxiliary lithium batteries utilizing a non-aqueous electrolyte, show higher vitality thickness than fluid electrolyte-based batteries because of the cell potential higher than 1.23 V, the thermodynamic restriction of water at 25°C. The magnificent exhibitions of non-aqueous lithium batteries may address the issue for high power batteries in miniaturized scale gadgets, compact hardware, and even electrical vehicles. Lithium does not exist as unadulterated metal in nature because of a high reactivity with air, nitrogen, and water. It is extricated from mineral or saline solution salt bog lithium chloride LiCl, lithium hydroxide LiOH, lithium carbonate Li_2CO_3. Since we just have 112.7 g of lithium in one kilo of Li_2CO_3, the extraction of one kilogram of lithium requires 5.3 kg of lithium carbonate. As indicated by the US Geological review (January 2010), Bolivia would house 32 % of world stores of lithium carbonate (Li_2CO_3), and Chile about 27 %. Be that as it may, Chile, China, and Argentina are the biggest makers. Note that 0.8 kg of lithium metal is delivered every second on the planet that is 25000 tons per year, for the most part to create lithium-particle batteries for electric autos or mobile phones. In the US Geologican overview, Goonan said that it would take 1.4–3.0 kg of lithium identical (7.5–16.0 kg of lithium carbonate) to help a 40-mile trip in an electric vehicle before requiring energize [15].

2.2. SECONDARY BATTERIES

In auxiliary batteries, the anode responses are reversible, i.e., applying the outside voltage to recreate the cathodes to their unique state, thus, the batteries can be rechargeable. Therefore, the rechargeable batteries go about as vitality source and also vitality stockpiling frameworks. The limit, rate ability, cycling execution, natural wellbeing and security issues are few of the huge execution attributes, which are worry in rechargeable batteries. A portion of the vital regular auxiliary batteries are quickly examined in this segment [16].

2.2.1. The Lead Acid Batteries

Lead-corrosive battery is outstanding and normal one among the accessible rechargeable batteries, which depends on the science of lead. In this battery, lead fills in as the anode and lead dioxide fills in as the cathode, which are dunk into an electrolyte arrangement of sulfuric corrosive [S]. Amid the release. the arrangement of uater changes the electrolytic action, which influences the open circuit voltage. The development of PbS04 amid the release caused the passivation of cathodes and decreases the pragmatic limit. Considering the restricted mass use and the need of inert mixes, for example, lattices, separators, cell holders, and so on., the viable estimation of particular vitality (WhiKg) is just – 25% of the hypothetical one. To upgrade the ionic conductivity in the charged and released states, an overabundance corrosive is important. Due to the constraints, for example, short life, high support and lacking vitality thickness, makers of rechargeable batteries are investigating different innovations.

2.2.2. The Nickel Cadmium Batteries

Nickel – Cadmium (Ni-Cd) is the principal little fixed rechargeable battery. In soluble (KOH) electrolyte, the Cd negative cathode works reversibly as indicated by an answer precipitation system with $Cd(OH)_2$ being the release item. The positive anode is $Ni(OH)_2$ which can reversibly de-embed 1 embed protons amid release 1 charge. The electrochemical responses of Ni-Cd battery, Ni-Cd batteries show long cycle life, high rates of accuse and release of steady release voltage and furthermore better low temperature execution. Nonetheless, the cost of the nickel – cadmium battery development, steadily chances related with the control of cadmium and furthermore the memory impact prompt the inquiry of interchange optional battery framework [17].

2.2.3. Nickel Metal Hydride (Ni-MH) Batteries

Nickel-Metal Hydride is a standout amongst the most exceptional financially accessible rechargeable frameworks and a naturally friendlier than the nickel-cadmium battery. The hydrogen stockpiling compound is a proton embedding's negative anode material that replaces the earth undermined cadmium negative cathode in the Ni-Cd hitter. The positive terminal is $Ni(OH)_2$ what's more, electrolyte is KOH. Ni-MH immediately traded the Ni-Cd for electronic applications as a result of its altogether higher vitality stockpiling ability and a broad research is moving to build up the multicomponent, multiphase composites for high hydrogen stockpiling limit, oxidation and erosion protection, quick gas recombination energy and lower cost. The significant issues to be comprehended for Ni-MH are poor low temperature ability, constrained high rate capacity and furthermore high self-release.

2.2.4. Secondary Lithium Batteries

Among the accessible optional battery frameworks, rechargeable lithium batteries a display most noteworthy particular vitality and vitality thickness. Another favorable position of lithium batteries is its most noteworthy individual cell potential over others. One extraordinary preferred standpoint of Li-particle batteries is their low self-release rate of just around 5% every month, contrasted and more than 30% every month and 20% every month individually in nickel – metal hydride batteries and nickel – cadmium batteries. Different parts of rechargeable lithium batteries are all around talked about in following segments.

2.2.5. Lithium Polymer Batteries

A Lithium polymer is in fact a lithium-particle polymer battery. It is fundamentally the same as the lithium particle battery however without a portion of the deficiencies. It can maintain a lot of mishandle. For instance, a completely charged Lithium Polymer battery can be punctured with a nail without blast or fire.

It initially utilized a plastic anode material and Solid Polymer Electrolyte (SPE) as the electrolyte. This relatively new innovation is quickly developing and moving forward. Right now, we utilize a gelled electrolyte and separator. The lithium-polymer electrochemistry at present covers an extensive variety of dynamic materials, for example, $LiCoO_2$, $LiNiO_2$, and its Co doped subsidiaries. Harding utilizes $LiCoO_2$ science.

As opposed to the customary metal can utilized by other little rechargeable cells, Lithium Polymer Batteries utilize a thin (110 μm), polymer-based bundling material to contain the electrochemical materials. This enables the framework to have a level thin (2 to 5 mm) shape factor. It is likewise conceivable to make the impression of the cell huge (e.g., 70 mm by 100 mm), this being in a perfect world suited to handheld gadgets, for example, PDA's.

Since the instance of the cell begins as a sheet of polymer-overlay, changing the impression of the cell is financially savvy. Additionally, if the Lithium Polymer cell utilizes a 'stacked' development, altering the cathode/electrolyte structure is likewise simple. Thusly, Lithium Polymer cells show adaptability in their mechanical properties and adaptability in their development.

2.2.6. Lithium Sulphur Batteries

Sulfur was presented as a positive cathode material without precedent for 1960s by Herbert and Ulam [14]. Since that time, very many endeavors have been attempted to build up the metal-sulfur batteries, some of them for the most part concentrating on essential Li/S cells. After 2008, extremely fast increment in the advancement of developing applications, for example, EV, military power supplies and stationary stockpiling frameworks for sustainable power source incited much higher interest for high performing batteries. Specifically, the expanding business sector of electric vehicles (EV) is by all accounts the most grounded inspiration for making Li/S batteries as the arrangement of decision for what's to come [18].

Sulfur offers one of the most elevated hypothetical particular limit (1675 mAh g^{-1}) among all current positive anode materials, which is a request of size higher than what is presently proposed by the change metal oxide cathodes. The framework could hypothetically convey vitality thickness of 2600 Wh kg^{-1}. In addition, sulfur is a rich component, non-harmful and amazingly shabby, which may radically diminish the last cost of the battery. A sensible focus of viable vitality thickness is in the scope of 400–600 Wh kg^{-1}, which is higher as contrasted and the established Li-particle cells, and which would permit to expand the scope of EVs car industry to 500 km.

There is positively most likely that Li/S batteries are considered as an extremely practical contender for the cutting-edge vitality stockpiling frameworks. Critical advances were made amid recent years, which are in a matter of seconds exhibited in additionally area. Be that as it may, there is as yet far to go and many difficulties to overcome, since the framework experiences a few uncertain downsides, which basically incite a substantial hole amongst reality and desires. In later piece of the talk, we likewise call attention to the principle impediments and confinements of the Li/S innovation. However, before that, it is important to portray the way the Li/S batteries works, as its unpredictable working system is altogether different from the established Li-particle cells.

2.2.7. Lithium Ion Batteries

The issue identified with the lithium metal (anode) prompt the innovation of lithium intercalated materials as negative cathodes. Contrasting with lithium metal, lithium intercalating mixes enhances the cycle life and in addition security yet brings down the phone voltage and the charge exchange rate. Amid the operation, lithium particles are carry between the cathodes through the electrolytes, while electrons are driven through the outside electrical circuit by the electrochemical potential, lithium particle battery, which is likewise called "armchair battery" with anode and cathode lithium addition materials. The cell design of the lithium-particle battery is given underneath, LiC_6(s) I Solvent, lithium salt I $LiCo0_2$(s).

An established Li-particle cell system comprises of both positive and negative anodes, which are isolated by an ionically directing electrolyte. Progress metal oxides, or all the more by and large the oxide mixes, are regularly utilized as dynamic materials for the positive cathode, and graphite for the negative one, where intercalation responses of Li^+ particles happen amid release and charge separately. The electrolyte is typically soaked up in a permeable separator, which goes about as a physical boundary maintaining a strategic distance from the contact between the cathodes, while taking into consideration Li^+ particles transportation. Amid release, Li^+ particles immediately course through the electrolyte from the negative cathode and intercalate into the positive one. Amid charge, the particles move the other way, spilling out of the positive cathode and intercalating to the negative anode [19].

Notwithstanding, the advancement of vitality stockpiling frameworks accomplished over the previous decades, the abilities of lithium battery advances are always being tested by the cutting edge multifunctional versatile gadgets, which are progressively requiring an ever-increasing number of exhibitions as far as vitality and power thickness. Vitality stockpiling frameworks were not ready to enhance with an indistinguishable rate from the electronic business advance. In this manner, elective cathode and anode materials with higher limits should be created. To defeat the charge stockpiling restrictions of addition compound cathodes, transformation materials, which experience electrochemical responses including more particles and electrons, are turning into a promising choice. Be that as it may, accomplishing the full exhibitions of these materials is as yet a test. With this viewpoint, two frameworks, lithium-oxygen (Li-O_2) and lithium/sulfur (Li/S), are under broad examinations. Li-O_2 cells experience the ill effects of numerous uncertain and prickly issues, and its commercialization is fairly visualized for the later years to come. Then again, Li/S framework is far route nearer to being discharged as a genuine item, and some new businesses or little organizations are as of now going for creating Li/S models for various applications.

Li/S innovation is required to offer 2–3 times the vitality thickness of the best performing Li-particle batteries, and decreased expenses. Nonetheless, numerous constraints are known for this framework, which prevent the total mechanical exchange to the business showcase. Next passage displays the essentials of the Li/S batteries, alongside its fundamental restrictions. The condition of-specialty of existing accomplishments is additionally quickly displayed [20].

2.2.8. Applications of Lithium Batteries

Lithium battery innovation discovered extensive variety of uses because of its upgraded execution, than other accessible vitality innovation and few of them are given underneath.

2.2.9. Mobile Electronics

Lithium particle rechargeable batteries are the best decision for a large number of versatile gadgets, for example, camera, portable PC, camcorder,

I-unit, and so on., on account of the light weight and upgraded over all execution.

2.2.9.1. Nanotechnology

Lithium battery innovation displays a promising application as a power source in the rising nanotechnology. Nano battery innovation is the current intriguing issue, which is basic for the developing Nano measure gadgets as power source [21].

2.2.9.2. Medical Tool Kits

Advances of lithium batteries, for example, higher particular vitality and vitality thickness make them potential contender for control source in implantable gadgets.

2.2.9.3. Space Applications

Space mission additionally requires batteries, which can furnish most extreme electrical vitality with least weight and volume. Since lithium batteries are especially fulfilling the above prerequisites, it is an enhanced and also interchange innovation for traditional batteries, which are broadly utilized as a part of room applications.

2.2.9.4. Military Applications

Lithium batteries are discovered extensive variety of uses in military supplies including interchanges, warm imaging, night vision, observation, compound recognition, look, and safeguard, undersea mines, etc.

2.2.9.5. Hybrid Vehicles

Lithium batteries are most encouraging for half and half electric vehicles because of its most elevated particular vitality, vitality thickness and have long cycle life [22].

CHAPTER 3

THERMODYNAMICS OF LITHIUM ION BATTERIES

CONTENTS

3.1. Overview 26

3.2. Accelerating Rate Calorimeter (ARC) 28

3.3. Rate Of Generation of Heat 30

3.1. OVERVIEW

Among the many sorts of rechargeable batteries, Li-particle batteries have high volumetric and the most elevated gravimetric vitality thickness, because of its littlest nuclear weight of 6.94 g/mol among every metallic component in spite of littlest thickness of 0.53 g/cm^3. They can meet numerous prerequisites for items, for example, electric vehicles and compact electronic gadgets. Quiet, Li^+/Li has an extremely negative standard diminishment capability of – 3.05V versus a standard hydrogen terminal (SHE).

A Li-particle battery comprises of a few electrochemical cells associated in parallel or potentially in arrangement to give an assigned limit or voltage. Each electrochemical cell has two terminals, the cathode and anode, isolated by an electrolyte that is electrically protecting yet conductive for Li^+ particles. Amid releasing, when a Li-particle battery fills in as a galvanic cell, Li^+ particles flow inside from the negative terminal (anode) to the positive (cathode), while electrons move remotely from the negative terminal to the positive anode. Amid charging, or when it acts as an electrolytic cell, Li^+ particles flow inside from the positive cathode to the negative anode, while electrons move remotely from the positive terminal to the negative terminal to look after charge [23].

Li optional batteries have been utilized for quite a long while to control purchaser gadgets and are generally being explored to control electric vehicles (EV's). An essential test for these batteries is security, under injurious and additionally ordinary working conditions. Amid battery charge/release, different concoction and electrochemical response and additionally transport forms occur. Some of these responses and procedures proceed additionally under open circuit conditions. They are to a great extent exothermic and may make warm amass inside the battery if warm exchange from the battery to the surroundings isn't adequate. This might be the situation if the battery is worked under protecting conditions or in a hot domain. It will make battery temperature rise fundamentally, so ‚hot spots' may frame inside the battery, subsequently gambling warm runaway.

Late leaps forward in nanotechnology have enhanced the power thickness and charge/release rates for Lithium particle (Li^+) cells. These new abilities empower applications that require high power thickness and high charge/release rates. These necessities are, obviously, notwithstanding the customary battery limits utilized as a part of versatile stimulation and convenient figuring applications. The powerful cells are perfect for use in control devices and other engine driving applications; they convey the

Li^+ points of interest in addition to the NiCd abilities, together with a low ecological effect (most sorts meet all requirements for transfer in the standard deny stream) [24].

As innovation makes accessible concentrated vitality gadgets (high vitality thickness) prepared to do quick conveyance high release rate and high-power thickness, security issues show up. As a kind of perspective point, a run of the mill Li^+ 4-cell, 2-Amperehour (Ah) battery pack stores a vitality level of around 100 kJ, while a hand projectile (150 g of TNT) has 600 kJ. The potential for individual damage and additionally property harm is clear if such measures of vitality are discharged in an uncontrolled way.

Battery security has a few measurements. Chief is to forestall individual damage or property harm. There is additionally the need to secure the gear that the battery powers. The battery itself must be secured, since it is a genuinely costly piece of the framework and can be troublesome or difficult to supplant on the off chance that it falls flat. The mechanical, electrical, and synthetic outline and fabricate of Li^+ batteries is a develop innovation with its security issues caught on. In late cases in which Li^+ battery disappointments made world media features, tainting of the synthetic segments appears to have been the underlying driver of the disappointments.

Because of a portion of the wellbeing concerns, battery merchants don't offer single cell Li^+ batteries to shoppers and even to most OEMs (aside from battery-pack producers). Rather, battery merchants offer multicell battery packs as OEM or customer items. These packs coordinate Li+ cells with complex security gadgets. Charging single, arbitrarily chose cells in arrangement or parallel blends without pack-style securities isn't prudent [25].

We will keep in mind the Gibbs free vitality, and skip its derivation here

$$dG = VdP - SdT + \sum_i \mu_i dn_i \qquad (1)$$

The last term is the substance potential. Generally, the synthetic potential is a measure of the collaboration vitality of the distinctive particles in a framework including the van der Waals and electrostatic associations between species. Scientifically, it is characterized as

$$\mu = \left[\frac{\partial G}{\partial n}\right]_{T,p} \qquad (2)$$

Similarly, as equity of temperature characterizes warm harmony, and fairness of weight characterizes mechanical balance, so balance of compound potential characterizes substance balance. For a battery in which the temperature and weight are settled, we can take a gander at the substance potential to look at the compound balance of the responses that are happening in the framework. We saw last time that it is the electrochemical responses that characterize the thermodynamics of the framework. Specifically, the meaning of the phone potential is given by

$$Fu = F\left(\phi^{\alpha} - \phi^{\alpha'}\right) = \mu_{e^-}^{\alpha} - \mu_{e^-}^{\alpha'} \qquad (3)$$

3.2. ACCELERATING RATE CALORIMETER (ARC)

A comprehension of the vitality discharge from synthetic responses and the potential for runaway responses is imperatively critical in the concoction business. At the point when the warmth produced by a substance procedure is more noteworthy than the conceivable warmth evacuation, the temperature will ascend with maybe calamitous impact. Most concoction responses continue with warm, and the temperature of beginning are vital parameters. Just in an adiabatic calorimeter can such a runaway response be dependably assessed by reenacting what can occur on a substantial scale by copying the most pessimistic scenario zero warmth misfortune conditions. The Accelerating Rate Calorimeter is the most surely understood and world's most broadly utilized adiabatic calorimeter and will give full data on the warmth and discharge indicating capability of a runaway happening. This will likewise empowered ideal conditions to be utilized to permit innately safe operation. Just when tests are led in a really adiabatic framework conceivable to scale-up from the research center [26].

The framework is contained in a metal circle 2.5 cm in distance across, commonly of Titanium or Hastelloy C. The example mass is generally 2–8 g yet would rely on the normal vitality discharge and kind of test compartment known as a bomb is utilized. The bomb is connected to the top segment on the calorimeter get together by a weight fitting and weight line prompts the weight transducer. A fine thermocouple is appended to the external surface of the bomb and the cover of the calorimeter at that point situated on the base area. The calorimeter has three separate warm zones. The best (top area) contains two radiators and a thermocouple, the side zone of the base segment contains four warmers and a thermocouple and the base zone at the base segment contains two radiators and a thermocouple. After set up and association, the calorimeter is fixed inside a blast confirmation control vessel. Subsequent to characterizing trial conditions on the PC, the test can initiate. The test conditions are a begin and end temperature and picking the extent of 'warm advances,' 'hold up time' and recognition affectability.' The framework will right off the bat warmth to the begin temperature. To do this, a little warmer in the calorimeter, the brilliant radiator, applies warm. This warms the specimen, bomb, and it's thermocouple. The calorimeter is cooler and this temperature distinction is seen by the three calorimeter thermocouples. The framework will then apply energy to the calorimeter warmers to limit the temperature distinction. This will proceed as the temperature ascends to the begin temperature. At the point when this begin temperature is achieved the framework will go into a hold up period, amid this time no warmth is given by the brilliant radiator. This permits the temperature contrasts inside the calorimeter to be lessened to zero. The calorimeter works adiabatically permitting the – calorimeter temperature to track test temperature. This hold up period (ordinarily 10–15 minutes) is trailed by a look for or seek period. Once more, amid this period (ordinarily 20 minutes) no warmth is given by the brilliant warmer, and any temperature float, upwards or downwards, is watched. On the off chance that there is upward temperature float this is caused by a self-warming response. The heat wait-look for strategy, the ordinary method of operation of the Accelerating Rate Calorimeter, will proceed until the point when an upward temperature float watches an exothermic response, more noteworthy than the chose affectability (typically 0.01–0.02°C/min). The framework naturally changes to the exothermic mode; the framework will apply warmth to the calorimeter coat to keep its temperature the same as the bomb/test [27].

The adiabatic control is the key component of the Accelerating Rate Calorimeter. The framework proceeds in the exotherm mode until the point when the rate of self-warming is not as much as the picked affectability, at this stage the warmth hold up look for technique resumes. At the point when the end temperature is come to (or an end weight is achieved) the test naturally stops and cooling, by compacted air, starts. The point of the Accelerating Rate Calorimeter is to finish the test to get a full time, temperature, and weight profile of the exothermic response in a protected and controlled way.

The ARC is appropriate to concentrate both the thermos kinetics and thermodynamics of artificially responsive exothermic frameworks. The ARC has various choices accessible, intended to consider batteries amid cycling and manhandle. The Gibbs free vitality (ΔG) of a framework might be resolved and utilized as a marker of danger potential. Besides, changes in ΔG in view of calorimetric and electrometric information might be utilized to examine the impacts of cycling on battery effectiveness. The ARC gives an adiabatic situation in which a specimen might be examined under states of unimportant warmth misfortune. The calorimeter design is to such an extent that the example is suspended inside an encased situation (a substantial obligation chamber with base and top) whose temperature is controlled amazingly decisively. In this way adiabaticity is accomplished by following the temperature of the test intently while working inside the ARC's dynamic range [28].

3.3. RATE OF GENERATION OF HEAT

The proposed strategy for estimation of warmth age rate depends on estimation of active warmth transition at the external surface of the assortment of intrigue. Figure 3.1 demonstrates a schematic of a warmth-producing barrel shaped cell with a warmth motion sensor appended on the external spiral surface.

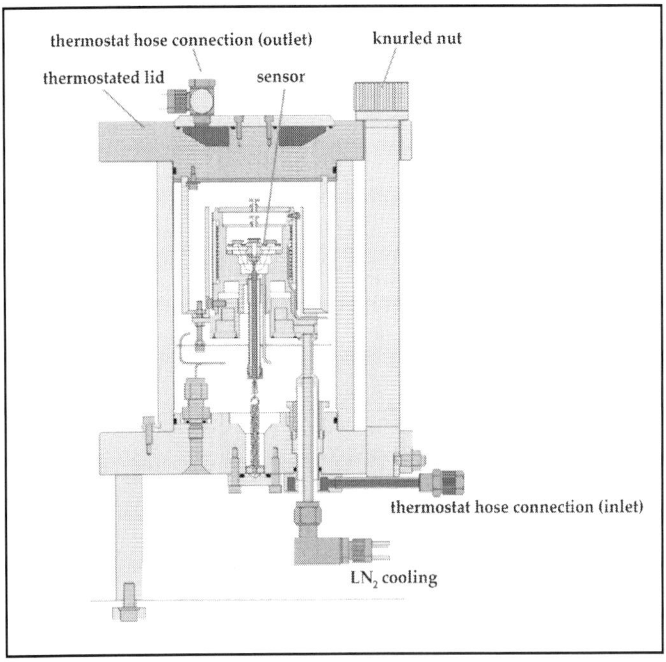

Figure 3.1: Schematic diagram of heat flux sensor. (Source: *https://dcyd0g-gl1hia3.cloudfront.net/fileadmin/_processed_/opt_Profile_of_DSC_204_HP_Phoenix_8c25d61c55.png?1439997090*).

Beginning at t = 0, the cell creates a uniform volumetric warmth age which is to be measured. Because of the volumetric warming, the warmth transition sensor measures an outward warmth motion that increments with time and in the long run achieves a consistent state esteem, $q''_{meas,SS}$. At enduring state, since the barrel temperature does not increment with time, all warmth created inside the phone is dispersed through the external surface. Accordingly, an estimation of the warmth transition sensor at enduring state can be utilized to decide the interior warmth age rate [29].

$$Q''' = q''_{meas,SS} \times \frac{A}{V} \tag{4}$$

where An and V are the cell external surface zone and volume individually.

On the off chance that the cell has been given adequate time to achieve warm enduring state, the deliberate warmth motion at consistent state can be utilized as a part of condition to decide the warmth age rate.

Be that as it may, as a rule, the operation of a Li-particle cell may not achieve warm relentless state. For instance, a Li-particle cell releasing at a high C-rate, say 10C, will experience finish release inside a couple of minutes. In such a case, it is vital to have the capacity to quantify the warmth age rate utilizing the warmth transition information accessible for just a constrained time, which might be much lower than the time expected to achieve consistent state. To empower such an estimation, a transient systematic model is created to display the deliberate warmth transition as an element of time. Accepting that the Biot number is little, the cell can be displayed as a lumped warm mass, and the temperature of the cell can be composed as

$$T_{cell}(t) = T_{cell,SS} \cdot \left[1 - \exp\left[-\frac{t}{\tau}\right]\right] \tag{5}$$

where τ is the warm time consistent.

Accordingly, the warmth transition through the external surface of the cell is given as an element of time by [30]

$$Q''(t) = q''_{meas,SS} \cdot \exp\left[-\frac{t}{\tau}\right] \tag{6}$$

Eq. (6) demonstrates that $q''_{meas,SS}$ can be resolved regardless of whether the cell does not really achieve warm enduring state. Eq. (6) has two obscure parameters $q''_{meas,SS}$ and τ. The warm time steady τ is by and large hard to decide from hypothetical computations, since it includes the convective warmth exchange coefficient to the surrounding medium, which isn't anything but difficult to quantify accurately. Notwithstanding, information from low power tests where $q''_{meas,SS}$ is known tentatively can be utilized to decide τ. This should be possible via completing a slightest squares filling of the information and deciding the estimation of τ that outcomes in minimum mistake. When τ is resolved from a low power analyze, for any investigation under similar conditions can be controlled by minimum squares attack of the information with above equation. This should be possible regardless of whether warm consistent state isn't really come to, since τ is known, and equation above has just a single obscure parameter to be resolved from slightest squares fitting. When that parameter is resolved, the warmth age rate can be acquired from condition.

A thoughtfully comparable approach includes the time subsidiary of the deliberate warmth transition variety with time. Separating condition and assessing at t=0 yields

$$\left|\frac{dq''}{dt}\right|_{t=0} = \frac{q''_{SS}}{\tau} \qquad (7)$$

Combining above equations results in

$$Q''' = \tau.\frac{A}{V}.\left|\frac{dq''}{dt}\right|_{t=0} \qquad (8)$$

Eq. (8) demonstrates that once the warm time consistent is known, the warmth age rate can be effectively decided from the underlying slant of the deliberate warmth transition information, regardless of whether the chamber does not really achieve warm enduring state. This equation kills the requirement for slightest squares fitting by utilizing the underlying slant of the warmth transition information. Be that as it may, since it is regularly hard to precisely decide the incline of exploratory information at a particular time, slightest squares fitting may in any case be beneficial.

Whereas the equations before it give the hypothetical premise to measuring heat age rates in a body regardless of whether the warmth age happens for a brief timeframe. The following area talks about investigations to approve this model.

Warmth creating rates inside the cell is gotten from the thermodynamic relations

$$Q = \Delta G + T\Delta S + W_{cl} \tag{9}$$

$$\Delta G = -nFE_{eq} \tag{10}$$

$$\Delta S = nF\frac{dE_{eq}}{dT} \tag{11}$$

$$W_{el} = -nF \tag{12}$$

$$q = I[(E_{eq} - E) + T\frac{dE_{eq}}{dT} + q_p \tag{13}$$

The thermodynamics of lithium-particle cells are convoluted by the nearness of fluid electrolyte blends and in addition single-stage and multiphase solids. Warmth age may come about because of blending and stage changes, and in addition the fundamental electrochemical responses. Solid forecast of temperature profiles of individual cells, and of a battery framework overall requires as a matter of first importance quantitative assessment of the aggregate warmth age rate. Wellsprings of warmth age:

1. The "reversible" warmth discharged (or retained) by the compound response of the cell.
2. The "irreversible" warmth age by ohmic protection and polarization.
3. The warmth age by "side responses," i.e., parasitic/consumption responses and "compound shorts."

CHAPTER 4

LITHIUM BATTERY STRUCTURE

CONTENTS

4.1. Construction of Battery 37

4.2. Electrodes Description 38

4.3. Anode 38

4.4. Cathode 39

4.5. Need Of Maintenance 43

Batteries work by changing over synthetic vitality into electrical vitality through electrochemical release responses. Batteries are made out of at least one cells, each containing a positive terminal, negative cathode, separator, and electrolyte. Cells can be isolated into two noteworthy classes: essential and optional. Essential cells are not rechargeable and must be supplanted once the reactants are exhausted. Auxiliary cells are rechargeable and require a DC charging source to reestablish reactants to their completely charged state. Cases of essential cells incorporate carbon-zinc (Leclanche or dry cell), antacid manganese, mercury zinc, silver-zinc, and lithium cells (e.g., lithium-manganese dioxide, lithium-sulfur dioxide, and lithium-thionyl chloride). Cases of optional cells incorporate lead-lead dioxide (lead-corrosive), nickel-cadmium, nickel-press, nickel-hydrogen, nickel-metal hydride, silver-zinc, silver-cadmium, and lithium-particle. For flying machine applications, auxiliary cells are the most conspicuous, yet essential cells are now and again utilized for controlling basic aeronautics hardware (e.g., flight information recorders) [31].

Batteries are evaluated as far as their ostensible voltage and ampere-hour limit. The voltage rating depends on the quantity of cells associated in arrangement and the ostensible voltage of every cell (2.0 V for lead acid and 1.2 V for nickel-cadmium). The most widely recognized voltage rating for air ship batteries is 24 V. A 24-V lead-corrosive battery contains 12 cells, while a 24-V nickel-cadmium battery contains either 19 or 20 cells (the U.S. military rates 19-cell batteries at 24 V). Voltage appraisals of 22.8, 25.2, and 26.4 V are likewise regular with nickel-cadmium batteries, comprising of 19, 20, or 22 cells, individually. Twelve-volt lead-corrosive batteries, comprising of six cells in arrangement, are likewise utilized as a part of many general flying airplane. The ampere-hour (Ah) limit accessible from a completely charged battery relies upon its temperature, rate of release, and age. Typically, airplane batteries are evaluated at room temperature (25°C), the C-rate (1-hour rate), and start of life. Military batteries, in any case, frequently are appraised as far as the end of life limit, i.e., the base limit before the battery is viewed as unserviceable. Limit evaluations of airplane batteries shift broadly, for the most part going from 3 to 65 Ah. The greatest power accessible from a battery relies upon its inner development. High rate cells, for instance, are outlined specifically to have low interior impedance as required for beginning turbine motors and assistant power units (APUs). Lamentably, no all-around acknowledged standard exists for defining the pinnacle control capacity of an air ship battery. For lead-corrosive batteries, the pinnacle control commonly is defined as far as the icy turning amperes, or CCA rating. For nickel-cadmium batteries, the pinnacle control rating regularly is defined regarding the current at greatest power, or Imp rating. These evaluations depend on various temperatures (18°C for CCA, 23°C for Imp), making it difficult to think about various battery sorts. Besides, neither rating satisfactorily portrays the battery's underlying pinnacle current ability, which is particularly imperative for motor begin applications. More thorough pinnacle control specifications have been incorporated into some military guidelines. For instance, MIL-B-8565/15 specifies the underlying pinnacle current, the current after 15 s, and the limit after 60 s, amid a 14-V consistent voltage release at two distinct temperatures (24 and 26°C). The condition of-charge of a battery is the level of its ability accessible in respect to the limit when it is completely charged. By this definition, a completely charged battery has a condition of-charge of 100% and a battery with 20% of its ability expelled has a condition of-charge of 80%. The condition of-strength of a battery is the level of its ability accessible when completely charged in respect to its evaluated limit. For instance, a battery appraised at 30 Ah, however just fit for conveying 24 Ah when completely charged, will have a condition of wellbeing of 24/30 100–80%. Along these lines, the condition of-wellbeing considers the loss of limit as the battery ages [32].

4.1. CONSTRUCTION OF BATTERY

Lead-corrosive cells are made out of substituting positive and negative plates, interleaved with single or different layers of separator material. Plates are made by gluing dynamic material onto a framework structure made of lead or lead composite. The electrolyte is a blend of sulfuric corrosive and water. In flooded cells, the separator material is permeable elastic, cellulose fiber, or microporous plastic. In recombinant cells with starved electrolyte innovation, a glass fiber tangle separator is utilized, some of the time with an additional layer of microporous polypropylene. Gell cells, the other kind of recombinant cell, are made by engrossing the electrolyte with silica gel that is layered between the cathodes and separators.

Lead-corrosive air ship batteries are built utilizing infusion formed, plastic monoblocs that contain a gathering of cells associated in arrangement. Monoblocs regularly are made of polypropylene, yet ABS is utilized by no less than one maker. Regularly, the monobloc fills in as the battery case, like a traditional car battery. For more powerful outlines, monoblocs are gathered into a different external holder made of steel, aluminum, or fiberglass-strengthened epoxy. Cases typically fuse an electrical repository for interfacing with the outer circuit with a fast associate/separate attachment. Two non-specific styles of containers are normal: the "Elcon style" and the "Gun style." The Elcon style is identical to military sort MS3509. The Cannon style has no military comparable, however is created by Cannon and other connector producers. Batteries now and then fuse thermostatically controlled warmers to enhance low temperature execution. The warmer is controlled by the airplane's AC or DC transport. Figure 4.1 demonstrates a gathering drawing of a run of the mill lead-corrosive flying machine battery; this specific illustration does not fuse a radiator.

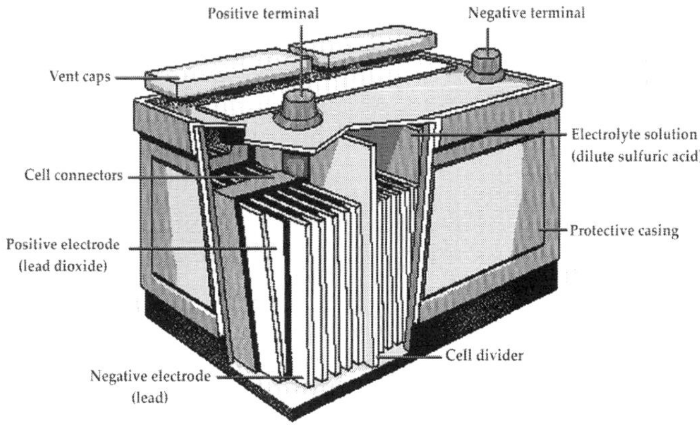

Figure 4.1: Drawing of lead acid battery. (Source: *http://www.reuk.co.uk//OtherImages/lead-acid-battery.gif*).

4.2. ELECTRODES DESCRIPTION

For a battery, not just the redox responses on both cathode and anode should be reversible, yet in addition the lithiation and de-lithiation happening in the two terminal materials must be for all intents and purposes reversible. Choosing materials for the two cathodes is to a great extent in light of the reversibility worry, and in addition thermodynamic soundness and electronic/ionic energy.

4.3. ANODE

The most punctual rechargeable Li batteries utilized lithium metal as the anode. In any case, a passivating layer was found to develop after rehashed working cycles of charge and release, and the metal surface wound up plainly uneven. This dendrite issue influences a battery to age rapidly and even detonate if the dendrite makes a short out between the anodes. Li-Al combinations enhanced wellbeing, yet at the same time couldn't keep the fast limit blur [33].

To date, carbon-based materials are the most widely recognized negative cathodes utilized as a part of Li-particle batteries. Graphite was observed to have the capacity to have Li in a steady stage with a stoichiometry of LiC_6. The graphite anode was marketed by Sony in 1991. The graphite anode has a hypothetical limit of 372 mAh/g, and just 0.1 V difference from a Li^+/Li terminal. Supplanting lithium metal with graphite takes care of the dendrite issue as well as brings down the cost. As of late, keeping in mind the end goal to enhance the lithium diffusion coefficient in carbon, graphene Nano circles and graphene oxide nanosheets were outlined with a higher surface-to-volume proportion. The fascinating electronic portability and mechanical properties of graphene are relied upon to help with the battery execution, however have not been verified yet. To push the point of confinement of vitality thickness and limit in carbon-base materials, other anode applicants are likewise being considered, including metals, composites, and oxides. For instance, the spinel $Li_4Ti_5O_2$ is being explored for high-control applications. Since its charging potential is 1.55 V higher than Li^+ /Li, it has no danger of lithium affidavit and is ok for quick cycling with a high present. Other than the possibility for lithium addition, there are additionally transformation sort materials that include more muddled concoction responses than Li intercalation into the host structure, including the silicon nanowires, silicon and germanium Nano crystallites, SnO_2 on carbon nanosheets, and TiO_2 on graphene substrate. All show great rate ability as anodes for lithium stockpiling, maybe in light of the short diffusion separations required for Li addition into nanostructures.

4.4. CATHODE

4.4.1. TiS_2 with Hexagonal Arrangement

As the most punctual cathode for lithium particle cells, TiS_2 has a layered hexagonal structure with open octahedral locales for reversible Li inclusion. It was first proposed by Exxon in 1972. This material structures a solitary stage with lithium fixation over the whole scope of $LixTiS_2$ for $0 \leq x \leq 1$.

4.4.2. $LiMPO_4$ with Olivine Arrangement

From that point forward, three noteworthy classes of inclusion sort cathodes have been produced. The most recent class is the lithium polyanionic salt. Among them, lithium metal phosphate, $LiMPO_4$, M = Fe, Mn, Co, Ni, with the olivine-sort structure, is the best known. The open-circuit voltages are 3.5 V for $LiFePO_4$, 4.1 V for $LiMnPO_4$, 4.8 V for $LiCoPO_4$, and 5.1 V for $LiNiPO_4$. Despite the fact that other change metal phosphates have higher possibilities than $LiFePO_4$, none indicated better electrochemical reversibility looked at than $LiFePO_4$. Definite data on $LiFePO_4$ and the olivine structure are given in the accompanying parts [34].

Different classes of polyanion-based cathode applicants incorporate $LiMPO_4F$ and Li_2MSiO_4 M=Fe, V, or Co. The lithium press fluoro phosphate ($LiFePO_4F$) has the same triclinic precious stone structure as tavorite $LiFePO_4OH$, in which each two Fe-focused octahedra (FeO_4F_2) share a F molecule, and each Fe-focused octahedron imparts four O iotas to four PO_4 tetrahedra around it. Li, in this structure, diffuses in three measurements rather than in a one-dimensional channel, as in $LiFePO_4$. $LiFePO_4F$ can be blended from processing FeF_3 and Li_3PO_4, or by an "ion-o-therma" response of blending similar antecedents in ionic fluids at controlled temperature. $LiFePO_4F$ demonstrated an electrochemical progress to Li_2FePO_4F, giving a limit near the hypothetical 152 mAh/g. The precious stone strain made by lithium intercalation is observed to be littler than $LiFePO_4$. The Li_2FeSiO_4 was accounted for to be reversible, with a limit of 140 mAh/g and around 3.0 V release voltage. Dissimilar to the olivine phosphates, Li_2FeSiO_4 gems comprise of FeO4 and SiO4 tetrahedra. The deliberate electrical conductivity is however 1000 times littler than that of $LiFePO_4$.

4.4.3. $LiM0_2$ with Hexagonal Arrangement

The class of cathode materials that has gotten the most broad investigation is the lithium metal oxides, $LiMO_2$ (M = Co, Mn, Ni, Al). Following the revelation of $LiCoO_2$ in 1980, its effective commercialization conveyed gigantic consideration regarding this class of material. Like TiS_2, $LixCoO_2$ has the hexagonal layered structure (O_3) for $0.75 \geq x \geq 0.5$, where Li can be facilitated between CoO_2 octahedra planes. At $x \approx 0.75$, a difference in electrical conductivity was watched, and opening are accepted to clutter. Underneath $x \approx 0.55$, the structure changes irreversibly to a monoclinic stage. On the off chance that all Li can be extricated, $LixCoO_2$ has a hypothetical limit of 274mAh/g. Nonetheless, just about portion of the limit is reversible in a down to earth detect. $LiNiO_2$ is difficult to combine with controlled stoichiometry, and for the most part winds up with $xLi_1 - yNi_1 + yO_2$. Additionally, Ni^{3+} is precarious with oxygen, causing control of unnecessary Ni particles on Li destinations and a high oxygen incomplete weight. These give $LiNiO_2$ the disservices of low lithium energy and poor security [35].

$LiMnO_2$ is more ecologically benevolent and prudent than $LiCoO_2$. However, not at all like CoO_2, the hexagonal precious stone structure of MnO_2 is difficult to shape by traditional strong state response, and the orthorhombic structure is observed to be all the more thermodynamically stable. $LiMnO_2$ is frequently acquired from trading Li particles with stable $NaMnO_2$. The metastable $Li0.5MnO_2$ regularly develops into a steadier spinel structure. Double and ternary frameworks with blended progress metals indicated engaging rate capacity and strength. Run of the mill illustrations incorporate $LiCo_1/3Ni_1/3Mn_1/3O_2$ and $LiNi0.5Mn0.5O_2$. Reversible limits of these blended metal mixes were accounted for to be 200 mAh/g over a potential scope of 2.5 to 4.5V versus Li^+/Li. Considerably less limit blur was watched contrasted with $LiCoO_2$.

As of late, different covering forms were utilized for enhancing electrochemical execution of the lithium metal oxides. This incorporates coatings of $AlPO_4$, SiO_2, and Al_2O_3 nanoparticles on the $LiMO_2$ to expand cycle life and decrease limit blur. The purpose behind such change is misty in any case. They may diminish the precious stone miniaturized scale strain of $LiMO_2$, or smother the exothermic response between the electrolyte and the terminal. The covering of Li_2MnO_3 was found to effectively balance out $LiMO_2$ cathodes, as the layered Li x $MnO_3 + (1-x)$ $LiMO_2$ structures to help accomplish a high limit (> 200 mAh/g) and high working voltage 3.0 V–4.5 V.

4.4.4. $LiMn_2O_4$ with Spinel Arrangement

Another option to $LiMnO_2$ is the spinel $LiMn_2O_4$ initially proposed in 1983. At the point when the Li particles move out, it moves toward becoming MnO_2 when charged as the cathode. The Mn^{4+}/Mn^{3+} particles with a capability of 4.1 V versus Li^+/Li give high solidness and power ability. The relationship between its cubic cross section parameter and lithium focus was researched. The most serious issue of these spinel cathodes is their ability blur amid cycling, particularly at hoisted temperatures with the electrolyte of LiPF6. Disintegration of Mn into the electrolyte is by and large thought about the primary driver. As of late, electrolyte added substances, fluorine substitution, and oxide covering were found to generally diminish Mn dissolution [36].

4.4.5. FeF_3 with Rhombohedral Arrangement

FeF_3, which crystalizes into the ReO_3 rhombohedral structure, is the fluoride cathode that has pulled in the most consideration by a long shot. An early investigation demonstrated an underlying release limit of 140 mAh/g took after by around 80 mAh/g after cycling. The release voltage was at 3.0V, 0.4V lower than the diminishment capability of FeF_3/$LiFeF_2$. Later work demonstrated that carbon covering by ball-processing FeF_3 with graphite or carbon nanotubes gave an expanded limit of 600 mAh/g. This demonstrates advance lessening past $Fe^{3+} \rightarrow Fe^{2+}$ at FeF_3. The response instrument was then examined by NMR and XRD thinks about, which found a change from ReO_3 structure to the rutile structure amid release, yet many points of interest stayed questionable. A few other iron mixes were considered throughout the years, for example, $KFeS_2$, FeS_2, $FePS_3$, FeOCl, however none indicated great reversibility.

4.4.6. Li-S and Li-O_2

For electric vehicle applications, Li-S, and Li-air batteries have gotten much consideration. Not at all like the significant three classifications of cathode materials that store charge by Li^+ addition (or extraction) in the host structure, at the Li-S anode the sulfur keeps being diminished by expanding lithium stoichiometry as "polysulfide" Li_2Sn atoms, n from 8 to 1. The Li-S terminal can achieve a limit as high as 1675 mAh/g, despite the fact that this is at just 2.1V versus Li^+/Li. Lamentably, this cathode has extremely poor electrical conductivity both as sulfur and Li_2Sn. There is likewise a huge hysteresis amongst charge and release, and the limit blurs quickly. Additionally, the terminal material lithium polysulfide was discovered broken up in the electrolyte. Carbon covering was accounted for to assist enormously with those deficiencies. As of late, sulfur was covered onto graphene nanosheets, and found to enhance cell cycle life and limit. Like the Li-S anode, a Li-air terminal uses O_2 to oxidize Li, framing Li_2O_2 on the cathode and giving 3.0V in respect to Li^+/Li. Reversibility of O_2 decrease was acknowledged by utilizing a non-watery electrolyte, however like the O_2 cathode in energy units, a great impetus is required for oxidizing Li_2O_2, and this impetus remains a test. Also, the framework suffers a voltage polarization of 1.5V in its cycling bend.

4.5. NEED OF MAINTENANCE

Routine support of lead-corrosive flying machine batteries is required to guarantee airworthiness and to expand benefit life. For vented-cell batteries, electrolyte topping must be performed all the time to recharge the water misfortune that happens amid charging. Upkeep interims are ordinarily 2 to 4 months. A limit test or load test as a rule is incorporated as a component of the adjusting system. For fixed cell batteries, water recharging clearly is pointless, yet intermittent limit estimations by and large are suggested. Limit check interims can be construct either with respect to date-book time e.g., each 3 to a half-year after the first year or working hours e.g., like clockwork after the first 600 hours. Allude to the maker's upkeep guidelines for specific batteries of intrigue [35].

4.5.1. Failures due to Non-Maintenance

The dominating disappointment methods of lead-corrosive cells are compressed as takes after:

- Shorts caused by development on the positive framework, shedding or mossing of dynamic material, or mechanical imperfections jutting from the matrix, showed by powerlessness of the battery to hold a charge (fast decrease in open circuit voltage).
- Loss of anode limit because of dynamic material shedding, extreme matrix erosion, sulfation, or passivation, showed by low limit as well as failure to hold voltage under load.
- Water misfortune and coming about cell dry-out because of releasing seal, rehashed cell inversions, or inordinate cheat (this mode applies to fixed cells or to vented cells that are disgracefully kept up), showed by low limit or potentially failure to hold voltage under load.

Location of these disappointment modes is clear if the battery can be expelled from the flying machine. The battery limit and load capacity can be measured straightforwardly and the capacity to hold a charge can be derived by checking the open circuit voltage after some time. Be that as it may, location of these disappointment modes while the battery is in benefit is more difficult. The more basic the battery is to the wellbeing of the flying machine, the more critical it moves toward becoming to identify battery blames precisely. Various on-board discovery plans have been produced for basic applications, basically for military flying machine [34].

Lead, the real constituent of the lead-corrosive battery, is a harmful (noxious) concoction. For whatever length of time that the lead stays inside the battery compartment, no wellbeing peril exists. Shameful transfer of spent batteries can bring about presentation to lead, notwithstanding. Natural directions in the U.S. what's more, abroad forbid the transfer of lead-corrosive batteries in landfills or incinerators. Luckily, a framework exists for reusing the lead from lead-corrosive batteries. Similar procedures used to reuse car batteries are utilized to reuse air ship batteries. Government, state, and neighborhood controls ought to be taken after for appropriate transfer methods [35].

CHAPTER 5

THERMODYNAMICS INVOLVED IN TRANSPORT PROCESS OF LITHIUM ION BATTERIES

CONTENTS

5.1. Introduction 46
5.2. Transport of Heat 48
5.3. Conditions of Interface 50
5.4. Set of Complete Equations 53

5.1. INTRODUCTION

Now if we want to understand about temperature appropriation and heat generation or the heat runways in Li particle battery cells is of awesome handy intrigue, we see that numerous corruption forms in Li particle batteries are caused by and a large portion of them are improved under temperature increment. Heat runaway may start as an exceptionally neighborhood wonder by starting an exothermal response at a nearby hotspot.

In this manner, the normal temperature of a cell isn't sufficient to decide the start of a warm runaway. A spatially settled scientific portrayal of the procedures on the size of the cell would permit to mimic the conduct of the cell under shifting working conditions and to distinguish the likelihood of perilous hotspots. Numerical demonstrating of Li-particle batteries on cell settled level was spearheaded by crafted by Newman and his associates and extended and refined by numerous different writers. The demonstrating approaches depend on transport conditions for Li particles and charges in the electrolyte and in the dynamic particles of cathode and anode under isothermal conditions. Most displaying approaches for the warm conduct of batteries are focusing on general warm adjust conditions for an entire cell, by consolidating phenomenologically thermodynamic contemplations on entropy or enthalpy changes inside a phone with sensible suspicions on out of balance forms like Joule warming, warmth of blending, Peltier effect and Soret effect. An approach in light of nearby coherence conditions for the temperature was introduced in, expanding past work on species and charge transport in batteries. These creators concentrated on determining perceptible conditions for the warmth transport in permeable cathodes utilizing the volume averaging strategy. The minute condition for the temperature was detailed without considering conceivable changes of the vehicle conditions portraying species and charge transport. Likewise, the interface conditions have not been inferred. Completely coupled models have been considered for battery stacks and different sorts of batteries. Additionally, in these cases the conditions were not inferred but rather detailed as adjust conditions in which referred to physical effects and also the response energy for all applicable synthetic responses were phenomenologically joined. A different way to deal with treat transport in electrochemical frameworks is the precise approach of non-equilibrium thermodynamics spearheaded via Landau. In this approach, just preservation conditions and general standards of measurable mechanics were utilized to determine the general type of the pertinent transport conditions. Just quantifiable amounts like thermodynamic subordinates and transport coefficients enter the conditions. The Onsager relations which ensure the energy of the entropy creation and force strict relations between the conditions for densities, charges, and temperature, have a focal influence in the inference. Substance responses can without much of a stretch be dealt with inside this formalism. The interface conditions can likewise be determined by essential progression requirements as e.g., protection of charges, vitality, and so forth and considering the physical and substance forms inside the interface. As opposed to the phenomenological approaches the terms adding to the warmth creation are deliberately inferred utilizing nothing else except for by and large legitimate thermodynamic relations and aggregate vitality preservation, rather than simply physical instinct and trial encounter. In this way it is verified that no wellspring of warmth is disregarded, if every hydrodynamic variable and all significant compound responses are considered. A particularly exquisite type of this formalism was as of late used to infer a hydrodynamic hypothesis of electromagnetic field in constant media, which can likewise be connected to charged and attractive fluids. Additionally, charge and species transport in Li particle batteries can be managed utilizing the non-equilibrium thermodynamics system. We will determine in the accompanying the total arrangement of conditions including all interface conditions for the particle, charge and warm transport in Li Ion battery cells. We will consider a spatial scale L which is vast contrasted with the scale LDL of the diffuse twofold layer (i.e., L LDL10–20 nm). Focused electrolytes are nonpartisan on this scale because of the solid Coulomb co-operations between the particles prompting a to a great degree little Debye screening length. Potential slopes are just in the twofold layer around the dynamic particles sufficiently solid to cause charge partition. Likewise, dynamic particles fulfill charge lack of bias despite the fact that the component for getting charge nonpartisanship is exceptionally different. In dynamic particles, it is the huge versatility of electrons which ensures charge impartiality. The charge of an embedded Li Ions is immediately protected by nearby revisions of electronic charges and the vehicle of electrons into the dynamic particles over the present gatherers. In the electrolyte charge impartiality will prompt very corresponded movement of negative and positive particles. We will make express utilization of charge lack of bias in the determination of our conditions. Notwithstanding normally lessening the quantities of important conditions, the utilization of charge nonpartisanship likewise ensures, that the vehicle coefficients showing up in principle are the quantifiable amounts of the electrolyte. Give us for instance a chance to consider the diffusion forms in a liquid salt comprising of positive and negative particles. All in all, the self-diffusion coefficients are different for the two sorts of particles. On the off chance that there were no association between the particles, we could hope to get free electric charges in unwinding forms regardless of whether the underlying irritation were unbiased due to the different diffusion length of the particles. In actuality, the solid Coulomb connection between the particles keeps the presence of free charges. The procedure pertinent for diffusion is in certainty the aggregate entombs diffusion process with a particularly defined bury diffusion coefficient for the two species. This exceptional amount for positive and negative particles will show up normally in our hypothesis. To infer the vehicle conditions in the electrolyte and the strong particles we utilize a general and thorough hypothesis for polarizable, directing media presented in and summed up in. The hypothesis depends on general thermodynamic standards, Maxwell's hypothesis for electromagnetic fields and Onsagers complementary relations. In spite of the fact that vehicle instruments in dynamic particles and electrolyte can be exceptionally different on the minute (nuclear) scale, their plainly visible frame only differ in the practical conditions and the span of the vehicle coefficients. For instance in dynamic particles charge is transported chiefly by unadulterated electronic conduction. The commitment of the particle diffusion in the dynamic particles to the electric current can be dismissed because of the huge versatility of the electrons contrasted with the particles. Charge transport in the electrolyte on the opposite side is solely because of ionic transport. Truth be told the move of electrons into the electrolyte would bring about the decrease of Li particles in the electrolyte to metallic Lithium and is thought to be one of the numerous corruption components in Li particle batteries. These gigantic differences in the tiny charge transport components show up perceptibly similarly as different significance of the particular charge conductivity coefficients. In electrolytes it is translated as focus subordinate ionic conductivity in the dynamic molecule we may utilize a steady electronic conductivity. Likewise the exchange number of particles in dynamic particles can be set to zero because of the vanishing commitment of particle diffusion to the electrical current [36].

5.2. TRANSPORT OF HEAT

The condition for the temperature can be gotten from the entropy adjust and the articulation for the entropy generation and the relation

$$T\partial_t s = c_p \rho \partial_t T + T \left.\frac{\partial s}{\partial c}\right|_T \partial_t c \tag{1}$$

where c_p and ρ are the specific warm per unit mass and the mass thickness of the electrolyte or the dynamic particles. Utilizing the thermodynamic connection ∂s ∂c|T = −∂μ ∂T|c and the coherence condition we obtain

$$c_p\rho\partial_t T = -\vec{\nabla}\vec{q} - \overrightarrow{N_+}\ \vec{\nabla}\mu + \vec{\mathrm{j}}\vec{\mathrm{E}} - T\left(\frac{\partial\mu}{\partial T}\right)\vec{\nabla}\overrightarrow{N_+} \tag{2}$$

Changing from the first warmth flux q to the renormalized warm flux Q we get

$$c_p\rho\partial_t T = -\vec{\nabla}\left(\vec{\mathrm{Q}} - T\overrightarrow{N_+}\left(\frac{\partial\mu}{\partial T}\right)\right) - \overrightarrow{N_+}\left(\frac{\partial\mu}{\partial T}\right)\vec{\nabla}c - \overrightarrow{N_+}\left(\frac{\partial\mu}{\partial T}\right)\vec{\nabla}T + \vec{\mathrm{j}}\vec{\mathrm{E}} - T\left(\frac{\partial\mu}{\partial T}\right)\vec{\nabla}\overrightarrow{N_+} \tag{3}$$

The third and keep going term on the correct hand side of Eq. are wiped out by the first one on the correct hand side, on the off chance that we disregard the spatial variety of the thermodynamic subsidiary. The temperature condition is consequently given by

$$c_p\rho\partial_t T = -\vec{\nabla}\vec{\mathrm{Q}} - \overrightarrow{N_+}\left(\frac{\partial\mu}{\partial T}\right)\vec{\nabla}c + \vec{\mathrm{j}}\vec{\mathrm{E}} \tag{4}$$

Using the constitutive relations equations can be transformed into

$$c_p\rho\partial_t T = -\vec{\nabla}\left(\lambda\vec{\nabla}T\right) + \frac{\vec{\mathrm{j}}^2}{ù} - T\vec{\nabla}\left(\beta\vec{\mathrm{j}}\right) + \left(\frac{\partial\mu}{\partial c}\right)\frac{\left(\overrightarrow{N_+} - \frac{t_+}{Fz_+}\vec{\mathrm{j}}\right)^2}{D} - T\vec{\nabla}\left(c\left(\frac{\partial\mu}{\partial c}\right)\frac{k_T}{T}\left(\overrightarrow{N_+} - \frac{t_+}{Fz_+}\vec{\mathrm{j}}\right)\right) \tag{5}$$

The temperature changes because of warm conduction and four different hotspots for warm. They are in the request of their appearance Joule's warmth, Thompson effect, warmth of blending and the Soret effect. The Thompson effect can likewise be composed as

$$T\vec{\nabla}\left(\beta\vec{\mathrm{j}}\right) = T\frac{\partial}{\partial}\ \vec{\mathrm{j}}\vec{\nabla}T := \mu\ \vec{\mathrm{j}}\vec{\nabla}T \tag{6}$$

Here we utilized charge protection and the Thompson connection for the Thompson coefficient μT = T ∂β ∂T. On the off chance that we disregard all commitments corresponding to kT the warmth condition lessens further to

$$c_p\rho\partial_t T = \vec{\nabla}\left(\lambda\vec{\nabla}T\right) + \frac{\vec{\mathrm{j}}^2}{\varkappa} - T\vec{\nabla}\left(\beta\vec{\mathrm{j}}\right) + \left(\frac{\partial\mu}{\partial c}\right)D\left(\vec{\nabla}c\right)^2 \tag{7}$$

The conditions for particle fixation, electric field particular electric potential defined by E =−∇φ and temperature can be connected in the electrolyte and additionally in the dynamic particles. The main difference is the electric conductivity, which is supplanted by the normal electronic conductivity σ of the strong dynamic molecule and carbon dark, which is typically used to upgrade the electronic conductivity of the terminals [37].

5.3. CONDITIONS OF INTERFACE

To couple the vehicle in electrolyte and dynamic particles we need to detail interface conditions, which depict the intercalation and deintercalation of particles.

5.3.1. Conditions for Current

The interface conditions depict the intercalation response and the deintercalation response individually on the mesoscopic scale i.e., past the size of the diffuse layer. For one stage responses, it is normally expected that the vehicle of particles over the interface is totally portrayed by the Butler Volmer approach. More confused responses may require the utilization of more detailed hypotheses, which can be effectively fused inside our hypothesis. Here we use for straightforwardness the Butler Volmer hypothesis. The present thickness over the interface is e because of the intercalation response is inside the Butler Volmer approach given as [38]

$$i_{sc} = i_0 \left(\exp\left[\frac{\alpha_a F}{RT} \eta_s \right] - \exp\left[\frac{-\alpha_c F}{RT} \eta_s \right] \right) \tag{8}$$

αA and αC with αA + αC = 1 are weighting the anodic and the cathodic commitment of the overpotential ηs to the general response. The over potential is the deviation of the electrochemical potential from the concoction balance between dynamic molecule and electrolyte. It is subsequently defined by

$$\eta_s := \Phi_s - \Phi_e - \frac{\mu_e - \mu_s}{z_+ F} \tag{9}$$

The overpotential vanishes clearly if the dynamic molecule is in balance with the electrolyte. Normally the overpotential is communicated with the assistance of the half-cell open circuit potential U_0 of the separate terminal in respect to a Li metal cathode. Without loss of sweeping statement setting $\Phi_{Li} = 0$, the open circuit potential U_0 can be composed as

$$U_0 = \mu_{Li} - \mu_s \tag{10}$$

Using this expression in equation above we get

$$\eta_s = \Phi_s - \Phi_e - U_0 - \frac{\mu_e - \mu_{Li}}{z_+ F} \tag{11}$$

Since the synthetic capability of the electrolyte μe is exceptionally different from the compound potential μLi of the Li metal terminal the last two terms in equation above don't cross out when all is said in done. Presenting the electrochemical capability of the electrolyte ϕe in respect to the substance capability of Li metal [39]

$$\varphi_e := \Phi_e + \frac{\mu_e - \mu_{Li}}{z_+ F} \tag{12}$$

we can rewrite it as

$$\eta_s := \Phi_s - \varphi_e - U_0 \tag{13}$$

what's more, get the standard type of the overpotential however with the electrochemical potential rather than the electric potential. The sufficiency i_0 in above equation is given by

$$i_0 = kc^{\alpha_a} c_s^{\alpha_a} \left(c_{s,max} - c_s\right)^{\alpha_c} \tag{14}$$

k is a response rate. c_s, max is the most extreme convergence of particles in the dynamic molecule. We expect that Li particles are not put away in the twofold layer that is, all Li particles are intercalated in the dynamic molecule or discharged into the electrolyte. There ought to likewise be no flux of negative charges over the twofold layer. Neither enter electrons the electrolyte nor intercalate negative particles from the electrolyte in the dynamic particles under perfect conditions. Both effects would prompt debasement, which could obviously be displayed by adjusting the interface conditions. The nonattendance of flux of negative charges over the interface particularly implies that the aggregate current over the electrolyte-molecule interface is because of transport of positive particles as it were. On the off chance that the molecule is totally filled i.e., c = c_s, max, it must be ensured by the interface conditions that no electrical current ~ j is conveyed by negative charge transporters over the interface. These conditions can be defined numerically in the accompanying path with the ordinary ~n pointing from the strong into the electrolyte [40]

$$\vec{j}_s \vec{n} = i_{se} \tag{15}$$

$$\vec{j}_e \vec{n} = i_{se} \tag{16}$$

$$\vec{N}_{+,s}\vec{n} = \frac{i_{se}}{z_+ F} \tag{17}$$

$$\vec{N}_{+,e}\vec{n} = \frac{i_{se}}{z_+ F} \tag{18}$$

5.3.2. Thermal Conditions

To determine the warm interface conditions we consider an infinitesimally extended little bit of the interface opposite to the ordinary n which focuses from the dynamic molecule into the electrolyte. The interface conditions are most advantageously inferred utilizing the shape of the warmth transport condition. We incorporate the temperature adjust condition over the infinitesimal little volume component, which contains the entire thickness of twofold layer. The term relative to ∇N+ does not add to the flux since the flux is monitored over the twofold layer. We get

$$\Delta dV c_p \rho \partial_t T = \iint dA(\vec{n}\left(\vec{q}_s - \vec{q}_e\right) + \vec{n}\vec{N}_+\left(\mu_s - \mu_e\right) + \vec{n}\vec{j}\left(\Phi_s - \Phi_e\right) \quad (19$$

Using Equations above and the definition of Q we obtain

$$\Delta dV c_p \rho \partial_t T = \iint dA(\vec{n}\left(\vec{Q}_s - \vec{Q}_e\right) + \frac{i_{se}}{z_+ F}\left(\mu_s - \mu_e + z_+ F\left(\Phi_s - \Phi_e\right)\right) - \frac{i_{se}}{z_+ F} T\left(\frac{\partial \mu_s}{\partial T} - \frac{\partial \mu_e}{\partial T}\right) \tag{20}$$

In the event that we expect that the twofold layer momentarily changes heaps of the streams, fixations, and temperatures, the twofold layers is dependably in a stationary state that is the left-hand side of above equation vanishes. Utilizing equations above we get

$$\vec{n}\left(\vec{Q}_s - \vec{Q}_e\right) = -i_{se} T \frac{\partial U_0}{\partial T} - \frac{i_{se}}{z_+ F} T \frac{\partial\left(\mu_e - \mu_{Li}\right)}{\partial T} \tag{21}$$

Disregarding the temperature reliance of the substance capability of Li metal and utilizing the constitutive connection for Q Eq. (21) we finally obtain [41]

$$-\lambda_s \vec{n}\vec{\nabla} T_s + \lambda_e \vec{n}\vec{\nabla} T_e = -i_{se}\eta_s - T i_{se}\left\{\left(\beta_s - \beta_e\right) + \frac{\partial\left(U_0 - \frac{\mu_e}{z_+ F}\right)}{\partial T}\right\} + i_{se}(c_s \frac{\partial U_0}{\partial T} k_{T,s} + c_e \frac{\partial \mu_e}{\partial T} \frac{k_{T,e}\left(1 - t_+\right)}{z_+ F} \tag{22}$$

Here we likewise utilized that the transference number of Li particles in the dynamic molecule might be dismissed because of the high versatility of electrons contrasted with the portability of Li particles. The physical significance of the different articulations on the correct hand side of equation is self-evident. The first one is the irreversible warmth generation because of Joule warming. The following term adds to the reversible Peltier effect and the last one is the Soret effect in the twofold layer. The Peltier coefficient is given as

$$Đ = T\left(\beta_s - \beta_e\right) + T\frac{\partial\left(U_0 - \frac{\mu_e}{z_+F}\right)}{\partial T} \tag{23}$$

Note that in Eq. (11) just the incomplete subordinate of the open circuit potential was given as Peltier coefficient. Our thorough approach demonstrates normally that additionally the differences in the Seebeck coefficients of the two stages and the warm subsidiary of the electrolytes synthetic potential are adding to the Peltier coefficient. Since our approach depends on tentatively available transport coefficients it is either conceivable to quantify straightforwardly the Peltier coefficient or to conclude it from estimations of the open circuit potential, the Seebeck coefficients of the two stages and the thermodynamic subordinates of the compound capability of the electrolyte.

5.4. SET OF COMPLETE EQUATIONS

5.4.1. For Transport in Electrolyte

The characteristic factors for the electrolyte are the particle fixation c, the electrical potential Φe and the temperature T. Because of the tentatively persuaded detailing of the interface conditions with the open circuit potential U_0 in respect to a Li metal terminal rather than the difference between concoction possibilities of strong molecule and electrolyte the electrochemical potential ϕ must be presented. It is along these lines more helpful to detail the vehicle conditions with ϕ_e rather t of Φ_e. One needs to remember that the electric field is as yet given by

$$\vec{E} = -\vec{\nabla}\Phi_e = -\vec{\nabla}\varphi_e + \frac{\vec{\nabla}\mu_e}{z_+F} \tag{24}$$

The transport equations are than given by

$$\partial_t c_e = \vec{\nabla}\left(D_e \vec{\nabla}\mu_e\right) - \vec{\nabla}\left(\frac{t_+}{z_+ F}\vec{j}\right) + \vec{\nabla}\left(\frac{Dc_e k_T}{T}\vec{\nabla}T\right) \tag{25}$$

$$0 = \vec{\nabla}\left(\varkappa \vec{\nabla}\varphi_e\right) - \vec{\nabla}\left(\varkappa \frac{1-t_+}{z_+ F}\left(\frac{\partial \mu}{\partial c}\right)\vec{\nabla}c_e\right) + \vec{\nabla}(\varkappa(\beta_e - 1/z_+ F\left(\frac{\partial \mu}{\partial T}\right) \tag{26}$$

$$c_{p,e}\rho\partial_t T = \vec{\nabla}\left(\lambda_c \vec{\nabla}T\right) + \frac{\vec{j}^2}{ù} - T\vec{\nabla}\left(\hat{a}_e \vec{j}\right) + \left(\frac{\partial \mu}{\partial c}\right)\frac{\vec{N}_+ - \frac{i_+}{z_+ F}}{D_e} - T\vec{\nabla}\left(c_e\left(\frac{\partial \mu}{\partial c}\right)\frac{k_{T,e}}{T}\left(\vec{N}_+ - \frac{t_+}{z_+ F}\vec{j}\right)\right) \tag{27}$$

Note that the first two conditions without the terms relative to the temperature slopes have the type of the conditions for the concentrated electrolyte by Newman, yet it is essential to understand that there exists a critical quantitative difference. In the concentrated electrolyte hypothesis of [4] the potential in Eq. (27) is deciphered as the electrical potential Φ_e. In our hypothesis, the electrical potential must be supplanted by the electrochemical potential ϕ_e of the electrolyte, on the off chance that we utilize the open circuit potential in respect to Li metal in the basic articulation for the overpotential [42].

5.4.2. For Transport of Active Particles

Our determination of the constitutive relations for fixation, electrical potential and temperature depends on general thermodynamic standards and can in this way be connected similarly to the vehicle in dynamic particles. Just if the vehicle is an isotropical the vehicle coefficients must be supplanted by tonsorial amounts. Here we confine ourselves to dynamic particles which act approximately isotropic on the micrometer scale. The conditions are somewhat simplified because of the sensible presumption that the transference number can be set to zero, since the electrical current is transcendently electronic of nature. We will signify the electronic conductivity with σ [43]. The vehicle conditions are then given by

$$\partial_t c_s = \vec{\nabla}\left(D_s \vec{\nabla}c_s\right) + \vec{\nabla}\left(\frac{Dc_s k_{T,s}}{T}\vec{\nabla}T\right) \tag{28}$$

$$0 = \vec{\nabla}\left(\sigma\vec{\nabla}\Phi_s\right) + \vec{\nabla}\left(\beta_s \sigma \vec{\nabla}T\right) \tag{29}$$

$$c_{p,s}\rho\partial_t T = \vec{\nabla}\left(\lambda_c \vec{\nabla}T\right) + \frac{\vec{j}^2}{\sigma} - T\vec{\nabla}\left(\hat{a}_s \vec{j}\right) - z_+ F\left(\frac{\partial U_0}{\partial c}\right)\frac{\vec{N}_+^2}{D_s} - Tz_+ F\vec{\nabla}\left(c_s\left(\frac{\partial U_0}{\partial c}\right)\frac{k_{T,s}}{T}\left(\vec{N}_+\right)\right) \tag{30}$$

Together with the interface conditions, the total arrangement of conditions for transport in Li particle batteries are figured. Obviously, the conditions must be shut by utilizing fitting limit conditions, however this is standard method. They may change from application to application contingent upon how the cell is coupled to the outside world. Particularly different cooling methodology will require different limit conditions either for the warm flux or the temperature itself. Likewise, the coupling of the present gatherer to the cathodes will influence the decision of limit conditions [44].

We have inferred the entire arrangement of transport conditions on a spatial scale bigger than the diffuse twofold layer in Li particle batteries. Our induction depends on general standards of non-equilibrium thermodynamics, which ensure thermodynamic consistency and particularly entirely positive entropy creation. The conditions for transport in the electrolyte differ from the concentrated electrolyte hypothesis. It can undoubtedly be said, that the concentrated electrolyte hypothesis of abuses the Onsager relations. In any case, we have likewise appeared over, that the coupled conditions for charge and species transport in Li particle batteries (disregarding heat transport) can be changed in the shape utilized, if in the electrolyte the unadulterated electrical potential is supplanted by the electrochemical potential in mix with the definition for the overpotential. The total arrangement of conditions can either be utilized to reproduce the vehicle inside cells utilizing a spatial portrayal of the cathodes, which settle the microstructure of the permeable terminals or as a beginning stage to acquire a permeable media portrayal of the anodes. In the minuscule approach, dynamic particles and electrolyte are dealt with as discrete media. So far, this ansatz is utilized for recreating the particle and charge transport under isothermal conditions. Incorporation of warmth transport is arranged as future work. The computational unpredictability of the issue enables just to reproduce delegate volume components (REV) of battery cells. Be that as it may, as post handling effective properties of the permeable anode might be acquired as e.g., effective diffusion coefficients or conductivities by numerically averaging over the REV. These amounts may then be utilized for reenacting the entire cell demonstrated with the permeable anode display. So far side responses as e.g., development of the strong electrolyte interface (SEI) or transport inside a SEI have not been considered. Inside our approach it is straight forward to incorporate those marvels. Surface responses will prompt a modification of the interface conditions. Transport inside a SEI requires the expansion of another thermodynamic stage with fitting transport coefficients. Volumetric responses require the consideration of the responding species and the data on the response energy. The induction of the vehicle conditions will continue along an indistinguishable line from appeared above with the main difference that the convergences of the species are not rationed independently but rather simply the whole of them. In future work we will make broad utilization of the model to acquire a superior comprehension of the coupled transport marvels and their repercussions on the execution and lifetime of Li particle batteries [45].

CHAPTER 6

THERMAL HEAT AND ENTROPY RELATED TO LITHIUM BATTERIES

CONTENTS

6.1. Thermal Heat Introduction 58
6.2. Entropy Introduction 60
6.3. Generation of Heat 62
6.4. Calculation of Heat.. 64
6.5. Thermal Runway of Battery For Its Safety 66

6.1. THERMAL HEAT INTRODUCTION

In thermodynamics, warm vitality alludes to the interior vitality introduce in a framework because of its temperature. The idea isn't all around characterized or extensively acknowledged in material science or thermodynamics, in light of the fact that the inside vitality can be changed without changing the temperature, and there is no real way to recognize which part of a framework's interior vitality is "warm." Warm vitality is once in a while utilized freely as an equivalent word for more thorough thermodynamic amounts, for example, the (whole) inside vitality of a framework; or for warm or sensible warmth which are characterized as sorts of exchange of vitality (similarly as work is another kind of exchange of vitality). Warmth and work rely upon the path in which a vitality exchange happened, while inward vitality is a property of the condition of a framework and would thus be able to be seen even without knowing how the vitality arrived [46].

In the laws of thermodynamics, Gibbs vitality is the most extreme conceivable non-development work yield done by a shut framework under a procedure with steady temperature and weight. In a shut electrochemical framework, non-extension work yield is electric vitality, so when the synthetic vitality is changed over in to electric vitality under a reversible procedure, the electric vitality rises to the Gibbs vitality, i.e., $\Delta G = -nFE$. At the point when the synthetic vitality is changed over into electric vitality under an irreversible procedure, the electric vitality is not as much as the Gibbs vitality, i.e., $nFE < -\Delta G$. The lingering Gibbs vitality is changed over into warm energy. In the second law of thermodynamics, entropy is a broad state work under a reversible procedure: $dS \equiv dq/T$, so under a reversible procedure, with steady temperature and pressure [47]

$$\Delta S = \int_1^2 \frac{dq}{T} = \frac{q}{T} = \frac{\Delta H}{T} = -\frac{\Delta G}{T} \tag{1}$$

furthermore, in a shut electrochemical framework, $\Delta S = -\Delta G/T = nF(E/T)$. At that point in a shut electrochemical framework, warm age $Q = T\Delta S = nFT(E/T)$ under a reversible procedure. In light of the laws of thermodynamics, the hypothetical capability of an electrochemical framework can be figured from the Gibbs vitality information, and the most extreme electric vitality that can be conveyed by the chemicals that are put away inside or provided to the cathodes in the cell relies upon the adjustment in Gibbs vitality ΔG of the electrochemical couple. The open capability of a cell can be acquired tentatively and is not exactly or equivalents to the hypothetical potential.

Both the hypothetical potential and the open potential are dictated by the kind of electro synthetic couples and the electrolyte contained in the cell. In a pragmatic cell, it is attractive that the majority of the Gibbs vitality can be changed over to helpful electric vitality amid discharge. Be that as it may, vitality misfortunes because of polarizations happen when a heap current goes through the cell, going with the electrochemical responses. The most vital factor that influences the vitality misfortunes of a cell is the polarizations. The aggregate polarizations of a phone include [48]:

(i) Ohmic polarization, which causes the voltage drop amid operation, and furthermore expends some portion of the helpful vitality as waste warmth. The aggregate ohmic polarization of a cell is the whole of the polarizations caused by the ionic protection of the electrolyte, the electronic protections of the cathodes, the present authorities and electrical tabs of the two anodes, and the contact protection between the dynamic materials and current gatherers. The Ohmic polarization takes after Ohm's law, with a straight connection between the current and the voltage drop.

(ii) Actuation polarization, which drives the electrochemical response at the anode/electrolyte interface, and

(iii) Focus polarization, which shows up because of the fixation contrasts between the reactants and the items at the cathode/ electrolyte interface and the fixation contrasts in the masses because of mass exchanging [49].

Every one of these polarizations cause utilization of Gibbs vitality, which is emitted as warmth vitality amid the charge-release process. In lithium cells, the dynamic materials are permeable, enabling lithium particles to be embedded in or separated from them amid the charge-release process, so the polarizations of a lithium cell are more mind boggling and the warmth age in lithium cells is likewise more intricate because of warmth being produced with each physical procedure. Lithium cells have high specific vitality, so the Gibbs vitality in cells is high. They will create high warmth vitality going with the change procedure from Gibbs vitality into helpful electric vitality. On the off chance that such warmth vitality can't be scatter the temperature of a shut lithium cell/battery expands, which may influence the execution of the cell/battery. Besides, due to the high Gibbs vitality in lithium cells, if side responses happen, more warmth vitality is changed over from the Gibbs vitality, expanding the temperature. At the point when the temperature of lithium cells is sufficiently high to actuate disintegration of terminals or electrolytes, a mishap may happen. Hence, the examinations of entropy in lithium cells have concentrated on assessing the warmth and evaluating the corruption of cells, since entropy is a broad state work. The examinations of warmth in lithium cells have concentrated on assessing and measuring the warmth. In view of assessing the warmth and measuring precisely, the administration and direction of lithium cells/batteries/frameworks could be worked easily and mishaps could be fewer [50].

6.2. ENTROPY INTRODUCTION

Entropy is essential for two fundamental reasons:

- Entropy is an essential idea in material science and data science, being the fundamental measure to look at changed conditions of a separated framework (the data substance of a portrayal). Entropy is the essential thermodynamic variable that serves to characterize and relate most warm properties of issue, and the balance state (for a segregated framework, the one with greatest entropy, what is known as the Second Law of thermodynamics).
- Entropy is engaged with the calculation of greatest and real efficiencies on most helpful building forms, what serves to classes the decency of acknowledge, and pinpoint where upgrades can be more beneficial.
- Entropy is a measure of the vulnerability on the condition of things (the motivation behind why everyone should wager at 7 in the two-dice amusement), a measure of how vitality and other broad amounts convey inside accessible requirements in a thermodynamic framework, etc., contingent upon the input acquired [51].
- We do not have a natural evaluation of entropy (in examination with vitality being identified with the speed and stature of a mass), yet we as a whole have a subjective sentiment entropy: everything has a tendency to get scattered, all movement tends to cease to exist, all articles tend to thermalize... Besides, there is more exact proof on the way that vitality tends to spread out, than on the way that vitality is moderate.
- Entropy (think as scattering), keeps an eye on a most extreme inside connected limitations. Here and there, most extreme entropy yields 'a uniform circulation' (e.g., thickness and arrangement in room air), yet frequently not (e.g., thickness and piece in a gravity field).

- By the way, entropy is the motivation behind why oxygen noticeable all around, which is heavier than nitrogen, does not tumble to the ground (the propensity to fall is exceeded with the inclination to scatter).
- Entropy estimation in Thermodynamics is about as basic as vitality estimation: vitality in a control-mass framework increments when work or warmth is included (thing on a cylinder barrel framework with caught air, and how work and warmth can be measured); framework entropy does not change when work is included 'easily,' and it increments in the sum dS=dQ/T when warm is included 'easily,' T being supreme temperature (whose zero is unattainable) [52].

As per the laws of thermodynamics, in a shut electrochemical framework $\Delta S = -\Delta G/T = nF(E/T)$, so the difference in entropy (ΔS) can be acquired through the incline of the open-circuit voltage (OCV) with temperature. The difference in entropy generally can be controlled by a potentiometric method [3]. In such a technique, the phone is released to a coveted condition of charge (SOC), and after an unwinding, the open circuit voltage goes to harmony, at that point the phone experiences a well ordered temperature variety, amid which the open circuit voltage is observed. Average consequences of the potentiometric technique incorporate a bend of the relating OCVs as elements of temperature and a line of the incline of the OCV versus temperature plot, as illustrated in Figure 6.1.

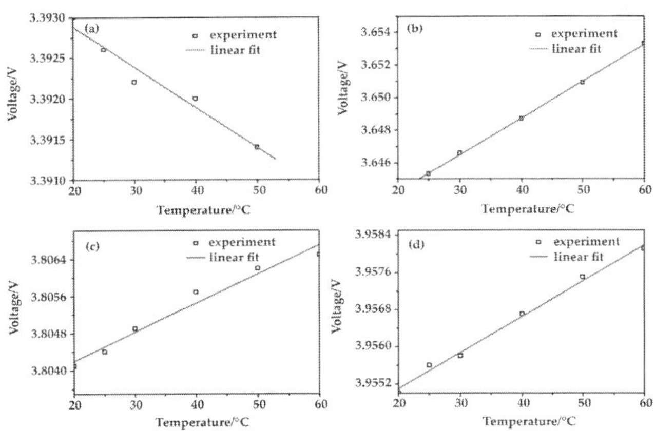

Figure 6.1: Graphs for OCV as a function of temperature for battery at different SOC (a) SOC = 0.122 (b) SOC = 0.458 (c) SOC = 0.644 (d) SOC = 0.813. (Source: Wang, S. (2016). Fundamental physics research in lithium batteries. Entropy and heat generation of lithium cells/batteries, 25(1).).

As of late, Schmidt et al. built up an electrothermal impedance spectroscopy strategy to decide the difference in entropy, in which the estimation times can be 100 times shorter than in a potentiometric method. The precision of this technique is like that of a potentiometric technique. In electrothermal impedance spectroscopy, the connection between warm flow inside the cell and the resultant temperature change can be misused, utilizing a sinusoidal current source. At the point when the warm exchange work (warm impedance) is known and the surface temperature is measured, the warmth flow inside the cell can be ascertained. The difference in entropy (ΔS) can be ascertained through the direct capacity between the warmth flow and increasing the current and the entropy. The ΔS in $LiFePO_4$ cells dictated by the regular potentiometric strategy and by the electro warm impedance spectroscopy have demonstrated comparative practices and are in great understanding. In any case, a hysteresis conduct of the ΔS has been seen, because of the superposition of the charging and release current.

6.3. GENERATION OF HEAT

The warmth age of lithium cells amid the charge and release process can be credited to two fundamental sources: the reversible warmth and the irreversible warmth.

6.3.1. The Irreversible Heat

The irreversible warmth is mind-boggling and is portrayed in various structures in various warmth assessing models [53].

6.3.2. The Reversible Heat

The reversible warmth is portrayed reliably as Qrev = TΔS = nFT(E/T) in all warmth assessing models. In the run of the mill electrochemical-warm model, the reversible warmth-creating rate is depicted as

$$q_{rev} = \sum_{j=n,p} a_{s,j} i_{n,j} T \frac{\partial E_j}{\partial, T'} \qquad (2)$$

whereas, j is the specific interfacial zone of a terminal, in, j is the superficial current thickness, E_j is open-circuit capability of a cathode response, N is negative anode, and p is certain cathode. In the commonplace proportional circuit warm model, the reversible warmth producing rate is portrayed as

$$q_{rev} = IT\frac{dE}{dT} \quad (3)$$

where I is the current. In this way, the reversible warmth producing rate can undoubtedly be figured from the difference in entropy or the difference in dE/dT, effortlessly.

6.3.3. Application

On the off chance that the conditions of cathode or electrochemical framework change, the entropy must change correspondingly, in light of the fact that entropy is a broad state work. In this manner, the difference in entropy can be connected in portraying the progressions of cathode structures and evaluating the condition of a cell/battery. Yazami et al. explored the entropy bend and the gem structure of lithium-intercalated graphite. The entropy bend demonstrates a sharp re-increment at x = 0.5 in LixC6, reacting to the progress from a very much requested stage-2 compound LiC^2_1 to an all-around requested stage-1 compound LiC_6, and the event of mediator phase(s) between the two lithium rich intercalation stages is confirmed by in situ XRD and the Raman spectra, amid lithium particle intercalating into graphite. In addition, the negative estimation of the entropy of intercalation at x > 0: 25 in $LixC_6$ is clarified by the vibrating recurrence of lithium molecules in graphite being higher than that in lithium metal. Lu et al. explored the progressions of entropy of $LiMn_2O_4$, Li1.156 Mn1.844 O_4, and Li1.06 Mn1.89 Al0.05 O_4 spinel cathode materials into equal parts cell systems. The outcomes demonstrate that the entropy profiles of the diverse spinel cathodes amid cycling associated well with the stage progress and the request/issue changes [54].

Besides, Mahera, and Yazamia built up a technique to evaluate the condition of debasement of lithium particle cells through the entropy and the thermodynamics conduct. They examined the impacts of cheat, cycle maturing and warm maturing on the entropy of lithium-particle batteries utilizing lithium cobalt oxide cathodes and graphite anodes. The entropy differs drastically with the connected cut-off voltage (4.2 V– 4.9 V). These progressions correspond well with the gem structure corruptions of the cathode and the anode. With expanding cycle number, the entropy indicates more significant changes than those saw in the release and the open-circuit potential bends particularly at specific conditions of charge and open-circuit potential esteems [55]. These distinctions are ascribed to the higher affectability of entropy state capacities to changes in the gem structure of the cathode and the anode instigated by cycle aging. Furthermore, the entropy indicates more evident changes with the maturing time than the open-circuit potential, when cells are put away at 60°C and 70°C. So, they recommend that entropy can be utilized to portray the debasement level of terminal materials and thus survey the phone's condition of wellbeing (SOH). Besides, Wu et al. propose that differential warm voltammetry (dT/dV) can be utilized for following debasement in lithium-particle batteries.

6.4. CALCULATION OF HEAT

The significant examinations of warmth age are the investigations on the procedures and components, notwithstanding the side responses (disintegration responses) in lithium cells and the warmth vitality changed over from the Gibbs vitality in each physical procedure and electro-concoction process.

Warmth assessment is important to deal with the warm conduct of the battery in scaled-up frameworks, and to enhance the efficiency of the cooling frameworks. Quantitative estimations and warmth figuring are helpful methods for assessing heat [56].

The quickened rate calorimeter (ARC), the warmth conduction calorimeter and the isothermal calorimeter have been utilized as a part of examinations of warmth age amid charge-release. In the ARC test, no warmth is lost to the environment, thus all the response vitality discharged concerns just the battery's self-warming. Then again, in both the warmth conduction calorimeter and the isothermal calorimeter, the warmth created amid charge-release is exchanged quantifiably. Quantitative estimations of warmth age of lithium cells are imperative for warm administration of scaled up battery frameworks.

Estimations of warmth amid charge-release are gotten however models of lithium cells/batteries. Among them, the proportional circuit-warm models and the electrochemical-warm models are the most widely recognized. In the comparable circuit-warm models, lithium cells are spoken to by circuits comprising of conventional electrical parts. The warmth created amid charge-release is isolated into reversible warmth (Q_{rev}) and irreversible warmth (Q_{irrev}). The reversible warmth (Q_{rev}) is ascertained by the difference in entropy (ΔS): $Q_{rev} = T\Delta S = nFT(E/T)$, as examining above. There are two normal techniques to compute the irreversible warmth (Q_{irrev}). One is ascertained through ohmic warmth: $Q_{irrev} = I^2R$, in which R changes with the changing of the conditions of cells, the operation and nature conditions, such as SOC, cycles, current thickness, temperature, etc. Another strategy is to figure through the vitality preservation and the voltage: Qirrev = nF (Ethe − E_{cur}), in which E_{the} is the hypothetical capability of the cell framework and E_{cur} is the genuine potential with current. The warmth estimations through the proportionate circuit-warm models are compact, so it has been utilized as a part of most warmth administration frameworks, and the precision of results relies upon the intricacy of the models. Choi et al. ascertained the warmth age of lithium particle cells utilized as a part of half and half electric vehicle (HEV) frameworks, with a specific end goal to build up a basic model to portray the warm conduct of an air-cooled Li-particle battery framework, proposed from a vehicle segment planner's purpose of view. Walker et al. figured the warmth age of lithium particle cells for space applications, and coupled it with particular orbital-warm programming, warm desktop (TD), to mimic the temperature versus profundity of-release (DOD) profiles and temperature ranges for all release and convection varieties with negligible deviation. Srinivasan et al. built up a model to figure warm age through five diverse inward parameters: the electrolyte protection (Rs), anode protection (Ra), cathode protection (Rc), and entropy changes in the cathode (ΔSc), and the anode (ΔSa). These five parameters are not reliant upon each other; they are subject to the condition of charge and the natural temperature. Hariharan built up a nonlinear comparable circuit display for lithium particle cells utilizing variable resistors, which are elements of the cell temperature. The model can be utilized to anticipate the cell voltage and temperature over an extensive variety of forces with a worldwide arrangement of parameters. In electro chemical warm models, the charge-release process is isolated into numerous physical and compound procedures, for instance, the dissemination of lithium particle in fluid and strong, the exchange of lithium amongst fluid and strong, the polarization on the surface of terminals, and so on. The warmth produced amid charge-release is the warmth impact of each physical and synthetic process, which usually can be computed as [57]

$$Q = S_a i_{loc}\left(\phi_1 - \phi_2 - E_{open}\right) + S_a i_{loc} T \frac{\partial E_{open}}{\partial T} + \kappa_1^{eff} \nabla \phi_1 . \nabla \phi_1 + \kappa_2^{eff} \nabla \phi_2 . \nabla \phi_2 + \frac{2RT\kappa_2^{eff}}{F}\left(1 + \frac{\partial \ln F}{\partial \ln c_2}\right)(1 + t_+)\nabla(\ln c_2) . \nabla \phi_2 \quad (4)$$

where E_{open}: open-circuit of the cathode; Sa: specific surface region of permeable area; i_{loc}: surface response rate; φ1: strong stage potential; φ^2: fluid stage potential; T: Kelvin temperature; κ_{eff} : powerful electronic conductivity of strong stage; κ^2_{eff}: successful ionic conductivity for fluid stage; R: perfect gas steady; F: Faraday's consistent; f: mean molar salt action coefficient; c_2: arrangement stage fixation; and t+: cationic transference number. The count of warmth created amid charge-release in light of electrochemical-warm models is extremely unpredictable, so it is utilized as a part of hypothetical research, yet not normally in applications. Kumares an et al. built up a warm model for $LiCoO_2$ or MCMB lithium particle cells to anticipate the release execution at various temperatures (15–45°C). Pals and Newman built up a one-dimensional warm model for a lithium/polymer cell to foresee the temperature profile in a Li/PEO15-$LiCF_3SO_3$ or TiS_2 cell stack release at 3 h rate. Baba et al. built up an improved single-molecule model to comprehend the warm conduct of lithium-particle cells and disperse data identified with nearby warmth age over the whole terminal plane, and a two-way electrochemical-warm coupled recreation technique has additionally been established.

6.5. THERMAL RUNWAY OF BATTERY FOR ITS SAFETY

In a mischance, the synthetic vitality in anodes may change over into warm vitality as opposed to electric vitality, which can actuate lithium cells into warm runaway. There are a couple of components that can lead lithium cells into warm runaway, among which the temperature of the lithium cell is one of the key determinants. The examinations of warmth age amid warm runaway can be utilized to anticipate the wellbeing and the criticality of lithium cells/batteries. The warmth age amid warm runaway can be measured by calorimeters that can bear the blast of lithium cells, for example, ARC. The estimations of warmth age amid warm runaway can get the execution of the warm runaway, firsthand. Feng et al. assessed the warm runaway highlights of a 25 Ah vast arrangement kaleidoscopic lithium particle battery with a Li (Ni_x Co_y Mn_z) O_2 (NCM) cathode by utilizing the broadened volume quickening rate calorimeter (EV-ARC). They found that it takes 15–40s from the sharp drop of voltage to the immediate ascent of temperature when warm runaway happens. Such a period interim can be utilized for early alarm of the warm runaway. The counts of warmth created amid warm runaway process are normally in light of the warm conduct of materials in the lithium cell. The consequences of figuring can be utilized to contemplate the cause and the impacts of warm runaway, with a specific end goal to enhance the wellbeing plan of lithium cells [58].

Richard et al. proposed a model for the warm runaway of a 18,650 carbon or Li^{+1} x Mn^{2-}x O_4 lithium-particle cell, in light of the warm security of de-intercalated Li^{+1} x Mn^{2-}x O_4 and lithium intercalated MCMB terminals in $LiPF_6$ EC: DEC electrolyte. The model has been utilized to anticipate the short out conduct and the broiler introduction conduct of the cell. The outcomes concur subjectively with that of trials. Kim et al. expanded the one-dimensional displaying approach defined by Hatchardetal to three measurements. The figuring aftereffects of broiler mishandle testing of cells with cobalt oxide cathode and graphite anode with LiPF6 electrolyte demonstrate that warm runaway will happen at some point or another than the lumped display, contingent upon the measure of the phone, and the responses at first spread in the azimuthal and longitudinal bearings to frame an empty barrel formed response zone. Wang et al. computed warm produced amid warm runaway of $LiFePO_4$ or C cells, and the outcomes demonstrate that the inward short out, caused by the softening down of the separator, is the central point of warm runaway of such cells, in which the separator with a drop liquefying down temperature has been utilized. Be that as it may, when the $LiFePO_4$ or C cell utilizes a separator with a higher dissolving down temperature, decay responses of cathode material turn into the main consideration of wellbeing.

CHAPTER 7

APPLICATIONS OF BATTERIES IN MODERN WORLD

CONTENTS

7.1. Aircraft..72

The world is waiting for the battery achievements in near future. About each area of the gadgets business, everything that keeps running on a battery, is constrained by the power yield and vitality life of the batteries that run it. Scientists are chipping away at enhancing the vitality thickness (juice per weight and volume), the value, the wellbeing, natural effect, and even the lifetime of the most famous class, lithium-particle batteries, and in addition planning entire new sorts [59].

Most batteries can be found in three principle enterprises: purchaser gadgets, car, and lattice stockpiling. "I would call them the three major cans of where individuals converge with batteries," says Venkat Srinivasan, appointee chief of innovative work at the Department of Energy's Joint Center for Energy Storage Research. Each pail has diverse prerequisites, and along these lines the batteries utilized can (in some cases) be altogether different from each other. That telephone in your pocket needs a battery that is smaller and safe, yet the weight and cost are less vital. Scale up to car batteries, and with such a significant number of batteries, cost and weight end up plainly critical, and in addition cycle life (you'd be extremely distraught if that new Tesla required new batteries each couple of years). Scale up significantly further, and the batteries that are starting to be utilized to store control for houses and the lattice have almost no weight or size prerequisites.

For a considerable length of time, shopper gadgets—your telephone, PC, camera, tablet, rambles, even your watch—have keep running on lithium-particle batteries, on account of their simple rechargeability and high vitality thickness. In these batteries, a cross section of graphite, loaded down with lithium particles, frames the anode. An oxide frames the cathode, associated with the contrary terminal, and the two are isolated by a fluid electrolyte that enables particles to go through it. At the point when the outer terminals are associated, the lithium oxidizes and the particles stream to the cathode. Charging is only the turn around. The more lithium particles that can be exchanged along these lines, the more power the battery can hold. We've come to welcome the reduced size and convenience, if not the battery life and security [60].

These long life batteries are utilized as a part of convenient shopper instruments like number crunchers, iPods, advanced journals, wrist watches and stop watches, toys, and simulated pacemakers. Lithium cells can likewise be utilized as a substitution of soluble batteries in numerous gadgets, for example, cameras and timekeepers. Despite the fact that they are more costly, lithium batteries will give any longer life. Silver Oxide batteries are utilized as a part of military and submarines. From antacid batteries to silver-oxide, the distinctive sorts of batteries are utilized as a part of various applications.

On account of autos, batteries are at last in charge of the lifetime of the auto and for the feared extend nervousness with regards to electric autos. To handle this issue, specialists, and researchers are endeavoring to pack more voltage limit into batteries. In any case, that is regularly connected with broken concoction responses, which lessen the limit after some time. A lot of research is given to finding new materials and chemicals to help or supplant the lithium-particle grid, or different parts of the battery [61].

Srinivasan brings up a couple potential developments, and these are not only for autos just: The customary graphite anode cross section could be supplanted with silicon, which holds 10 fold the number of lithium particles. Yet, silicon has a tendency to extend as it ingests lithium, so batteries should represent that. Or then again: Instead of the grid, lithium metal could go about as the anode—if we can make sense of how to keep it from calamitously shorting out when it is energized. It's an issue that battery makers have been attempting to understand since the lithium-particle battery was developed decades prior [62].

This exploration will be compelling to the batteries in our pockets and vehicles. Be that as it may, there's a third classification, where the effects are worldwide.

Melanie Sanford is utilizing demonstrating apparatuses on an alternate kind of battery—gigantic, redox stream batteries that will store control from inexhaustible power plants and discharge it when the breeze and sun aren't accessible. Night out the pinnacles and valleys of vitality creation and utilization will cause renewables scale up to give something other than supplementary power.

Southern California Edison is as of now trying different things with battery banks, utilizing Tesla auto batteries, but since the batteries are conventional lithium particle based, they're excessively costly, making it impossible to use on a scale that will permit worldwide inexhaustible power. In addition, the limitations for a framework battery are entirely different than an auto. Weight and size are not an issue, but rather cost and lifetime are.

Planning a battery for another air ship application or for retrofit requires a cautious frameworks building approach. To work well, the battery must be interfaced deliberately with the airplane's electrical framework. The battery's dependability and practicality depends intensely on the kind of charging framework to which it is associated; there is a fine line amongst undercharging and cheating the battery. Numerous airframe producers have understood that it is smarter to get ready specifications for a "battery framework" as opposed to having separate specifications for the battery and the charger. This approach guarantees that the charging profile is tuned accurately to the specific attributes of the battery and to the flying machine's operational prerequisites [63].

7.1. AIRCRAFT

As a rule, the airplane battery must be estimated to give sufficient crisis energy to help flight fundamental loads in case of disappointment of the essential power framework. FAA directions force a base crisis control necessity of 30 min on every single business plane. A few carriers force a more extended crisis prerequisite, for example, 40 or 60 min because of regular awful climate on their courses or for different reasons. The crisis prerequisite for Extended Twin Operation (ETOPS) forced on two-motor flying machine working over water is an entire 90 min, in spite of the fact that 60 min is permitted with working confinements. The specified crisis control prerequisite might be satisfied by batteries or other reinforcement control sources, for example, a slam air turbine. In the event that a slam air turbine is utilized, a battery still is required for transient fill-in. Specific necessities relating to air ship batteries can be found in the Federal Aviation Regulations (FAR).

For global applications, Civil Aviation Authority (CAA) and Joint Airworthiness Authority (JAA) controls ought to be counseled for extra necessities. At the point when utilized for APU or motor beginning, the battery must be estimated to convey short blasts of high power, instead of the lower rates required for crisis loads. APU begin prerequisites on substantial business air ship can be especially requesting; for example, the APU utilized on the Boeing 757 and 767 planes has a pinnacle current necessity of 1200 A [Gross, 1991]. The heap on the battery begins high to convey the in-surge current to the engine, at that point falls quickly as the engine creates back electromotive power (EMF). Inside 30 to 60 s, the heap drops to zero as the APU lights and the starter cutoff point is come to. The most pessimistic scenario condition is beginning at elevation with a cool APU and an icy battery; typically, a lower temperature farthest point of – 18°C is utilized as a plan point. A thorough plan technique for streamlining air ship starter batteries was produced by Evjen and Miller [1971]. At the point when nickel-cadmium batteries are utilized for APU or motor beginning applications, FAA directions require the battery to be secured against overheating. Appropriate means must be given to detect the battery temperature and to detach the battery from the charging source if the battery overheats. This prerequisite began in light of various occasions of battery warm runaway, which typically happened when 19-cell batteries were charged from the 28-volt DC transport. Most cases of warm runaway were caused by debasement of the cellophane gas boundary, accordingly permitting gas recombination and resultant cell warming amid charging. Present day separator materials (e.g., Celgard) have enormously lessened the event of warm runaway as a disappointment method of nickel-cadmium batteries, yet the likelihood still exists if the electrolyte level isn't appropriately kept up [64].

PART II
FUEL CELLS

CHAPTER 8

OVERVIEW

CONTENTS

8.1. Introduction 78
8.2. Emergence 80
8.3. Problems In Emergence of Fuel Cell 83
8.4. Difference Between Fuel Cell And Battery 84

The second part manages energy units, which has significance in both transport and stationary power applications. Once more, this begins with thermodynamic parts of the gadget operation and essential definitions of segments and ideas found in power modules. The resulting areas at that point go into assist clarification of the distinctive sorts: proton trade film energy component (PEMFC), coordinate methanol power device (DMFC), liquid carbonate energy component (MCFC), strong oxide fuel cell (SOFC) and soluble power module (AFC). With PEMFC, which are the most significant for car utilizes, a recorded diagram is given covering the improvement of the innovation yet it additionally addresses building difficulties, for example, warm administration, chilly begin and water administration. The advance and confusions of each energy component innovation at that point takes after with a short review of phosphoric corrosive (PAFC), coordinate carbon power device (DCFC), bacterial power device (BFC) and redox flow battery (RFB). RFB specifically are accepting much consideration in the huge scale vitality stockpiling industry because of their potential cost investment funds and expanded lifetime contrasted with batteries with the normal all-vanadium and iron-chromium RFB examined [65]. The area finishes up with discourses about power module applications and the generally introduced energy unit units for stationary apparatuses are outlined. Just PEMFC has gotten significant mechanical consideration for car utilizes and the creators propose the underlying additions are in overwhelming merchandise vehicles. Their innovation standpoint underlines their understanding of the anticipated issues which are growing new impetuses for oxygen lessening, impetuses for the total oxidation of ethanol and particular impetuses which don’t advance the undesired side responses.

8.1. INTRODUCTION

Energy components (FC) are electrochemical motors. Their part for natural insurance can’t be belittled. The fundamental favorable position of energy units in contrast with warm motors is that their efficiency isn’t a noteworthy capacity of gadget estimate. A power device consistently changes a piece of compound vitality into electrical vitality by devouring fuel and oxidant. In the present paper control amplification approach is connected to determine control confines in defective energy units, where for sufficiently vast electric streams control diminishes with current in light of winning impact of misfortune wonders In a past work [Sieniutycz, 2003] we displayed influence creation and its points of confinement in simply warm frameworks with finite rates.

Specifically, radiation motors were broke down as nonlinear frameworks administered by laws of thermodynamics and transport. Temperatures T of asset media were just fundamental factors to depict these frameworks. Be that as it may, energy units are more broad frameworks in which the two temperatures T and concoction possibilities mk are fundamental. They are electrochemicalenginespropelledbyfluxesofbothenergyandsubstances. Essential structure of power devices incorporates electrolyte layer in contact with a permeable anode and cathode on either side. Vaporous powers are bolstered constantly to the anode (negative terminal) compartment and an oxidant (i.e., oxygen from air) is encouraged persistently to the cathode (positive anode) compartment. Electrochemical responses happen at the anodes to deliver an electric current. Essential response is the electrochemical oxidation of fuel, generally hydrogen, and the lessening of the oxidant, typically oxygen. This guideline makes a power module like a compound motor. In a FC procedure in Figure 8.1 surges of fuel (H2) and oxidant (O2) interface; the procedure is impelled by diffusive or potentially convective fluxes of warmth and mass, exchanged through the cell 'conductances' or limit layers. The vitality flux (control) is made in the cell generator which misuses the fuel stream reaching with the anode and the oxidant stream reaching with the cathode. The two anodes are isolated by the electrolyte. As in normal warm machines and radiation motors both exchange instruments and properties of leading layers influence the rate of energy creation. Energy component frameworks working in the power yield mode are electrochemical flow motors impelled by concoction responses. Their execution is dictated by sizes and headings of partaking streams and by system of electric current age. Voltage bringing down in a phone underneath its reversible esteem is a decent measure of the phone flaw. The objectives of the present paper include [66]:

(a) detailing of a thermo-electro-compound model for flawed energy components, particularly for those with inadequate substance transformations.

(b) execution of the model to reenact the conduct of high-T strong oxide energy components.

(c) expectation of different misfortunes of the voltage and their impact on the cell execution.

(d) utilization of energy component attributes to determine influence limits.

The essential thermodynamic and electrochemical standards behind energy component operation and innovation are portrayed. The essential electrochemistry standards deciding the operation of the power module, the energy of redox responses amid the energy component operation, the mass and vitality transport in an energy component, and so on., are portrayed quickly to give a comprehension of down to earth energy unit frameworks. The perfect and reasonable operation of energy units and their productivity are likewise portrayed. This will give the system to comprehend the electrochemical and thermodynamic fundamentals of the operation of power modules and how energy unit execution can be impacted by the working conditions. The impact of thermodynamic factors like weight, temperature, and gas focus, and so on., on power device execution must be dissected and comprehended to foresee how energy units associate with the frameworks where it is connected. Understanding the effect of these factors permits framework investigation investigations of a particular energy component application.

8.2. EMERGENCE

Advancement in human culture, and particularly present-day progress, has been set apart by regularly expanding vitality utilization and power necessities. Most of the vitality needs have been given by ignition of petroleum products since the modern upset. Warmth motors using non-renewable energy source burning have brought about extreme neighborhood air contamination, debilitating the wellbeing of a large number of individuals living in huge numbers of the world's urban zones. They keep on contributing fundamentally to the expansion in the climatic carbon dioxide fixations, in this manner heighten the possibility of a worldwide temperature alteration. Notwithstanding the wellbeing and natural concerns, an unfaltering consumption of the world's constrained petroleum derivative stores and the very survival of mankind call for new age innovation for vitality change and power age, which is more effective than the customary warmth motors with insignificant or no contamination outflows, and furthermore perfect with sustainable power sources and bearers for economic advancement and vitality security. Power module has been distinguished as the most encouraging and potential vitality transformation innovation, which meets the greater part of the above necessities. Indeed, power device innovation has been effectively utilized as a part of numerous particular zones, strikingly in space investigations, where power module works on unadulterated hydrogen and oxygen with more than 70% warm to electrical vitality productivity and the main result water constitutes the sole wellspring of drinking water for the team of the shuttle. There are currently a few hundred energies component units for earthbound applications, from stationary cogeneration, versatile transportation to convenient applications, working in finished twelve of nations, amazing specialized advance has been accomplished, and is driving the improvement of aggressively valued power module based power age frameworks with cutting edge highlights [66].

Other than being effective, perfect and good with future vitality sources and transporters, energy unit likewise offers numerous extra favorable circumstances for both portable and stationary applications. Energy unit is an electrochemical gadget and has no moving parts with the exception of fringe compressors and engines. Accordingly, its operation is tranquil, and for all intents and purposes without vibration and commotion, along these lines equipped for being sited at the premises of the buyer to wipe out power transmission lines. Its characteristic particularity takes into consideration straightforward development and operation with conceivable applications for scattered, circulated, and convenient power age, since it might be made in any size from a couple of watts to megawatt scale plant with measure up to productivity. Its quick reaction to the changing burden condition while keeping up high productivity makes it in a perfect world suited to stack following applications. Its high effectiveness speaks to less substance, warm and carbon dioxide outflows for a similar measure of vitality transformation and power age. At display, power device is being utilized routinely in space applications, and has been under escalated advancement for earthly utilize, for example, for utilities and zero emanation vehicles. There exist an assortment of energy units, and they can be characterized in view of their working temperature, for example, low and high temperature power devices, the sort of particle moving through the electrolyte, and so on. Be that as it may, the decision of electrolyte characterizes the properties of an energy unit. Consequently, power module is regularly named by the idea of the electrolyte utilized. There are by and by six noteworthy power device innovations at different phases of advancement and commercialization. They are soluble, phosphoric corrosive, polymer electrolyte layer, liquid carbonate, strong oxide and direct methanol power modules. Their electrochemical responses, operation essentials, development and plan, application and best in class innovation can be discovered somewhere else. Albeit various examinations going for creating energy unit innovation as a reasonable wellspring of energy have been directed, some perplexity and misguided judgment exists about the thermodynamic execution of power modules and its correlation with warm motors. For illustration, it is regularly said that energy units are not constrained by the Carnot effectiveness, in this manner they have, or can accomplish, higher vitality proficiency than the customary warmth motors. Some even go so far as to express that power modules are not subject to "thermodynamic constraints." It is additionally regularly the training that in demonstrating an energy unit has better execution, a reversible energy unit is contrasted and the irreversible warmth motor absolutely an uncalled-for examination since a reversible vitality transformation framework is guaranteed to have preferable execution over the irreversible vitality framework in view of the second law of thermodynamics. Further, many examinations utilized the particular liquids included, and accepted steady properties, leaving to questions with respect to the all-inclusive statement of the conclusions came to. Besides, a few investigations even demonstrate the likelihood of power devices that could accomplish more than 100% effectiveness – an outcome some even refer to as a proof that energy unit is more proficient than traditional warmth motor since warm motor isn't conceivable to accomplish more than 100% productivity. Then again, Li just uses the thermodynamic laws without thought of particular liquids included and their property varieties, and exhibit that both power devices and warmth motors have a similar most extreme conceivable effectiveness while working on a similar fuel and oxidant, and this greatest proficiency and the Carnot productivity are equivalent. In any case, viable restrictions (irreversibilities) result in power modules having higher down to earth productivity than warm motors. Remarkable power device operation at much lower temperatures additionally yields less or no contamination emanations, accordingly prompting ecologically clean vitality change. At last, the conundrum about the likelihood of more than 100% vitality change proficiency for some reversible power modules was illuminated. Li's approach has likewise been utilized by others in comparative investigations [67].

In this part, the general approach by Li will be followed in the examination of energy unit execution. The strategy for examination will be general, without considering particular working liquids included or their particular varieties of properties with temperature, weight, and focus, so the outcomes acquired and conclusions came to will be for the most part pertinent. In particular, the reversible cell potential will be determined by utilizing the first and second laws of thermodynamics, and its varieties with the working conditions, for example, temperature, weight, and reactant fixations in the reactant streams will be gotten with general thermodynamic relations. The issue of vitality change productivity will be given the assistance of the first and second laws of thermodynamics. The most extreme conceivable effectiveness for power devices will be explored; an examination will be made with Carnot proficiency, which is the greatest conceivable productivity for warm motors against which energy units are going after business achievement. At that point, the likelihood of more than 100% proficiency for power devices is inspected. The vitality change productivity for an energy unit framework containing power modules and assistant types of gear will be considered, and effectiveness misfortune system for working power devices will be talked about. It is the goal of setting up this section the thermodynamic execution of energy unit and its correlation with its adversary warm motor is legitimately managed so any future mistaken assumptions and confusions can be maintained a strategically be ignored.

8.3. PROBLEMS IN EMERGENCE OF FUEL CELL

Power module innovation is settled in specialty advertise (military, aviation) since the second 50% of nineteenth century, while the commercialization in a bigger market is going gradually. Energy component frameworks have been a work in progress for quite a few years and their materialness has been shown overall both in little and in vast scale. By the by, effective commercialization of this innovation requires to conquer a few obstructions. Presently the principle challenges identified with the dispersion of energy units are low costs, quality, and acknowledgment by end-clients. Expenses are identified with the hydrogen creation and fabricating and depend likewise by Government's backings. Specialized issues are for the most part connected to the vigor, dependability, and solidness of a power device particularly contrasted with inward burning or turbine motors. Luckily, general society acknowledgment of power module could be less demanding considering that it is a domain well disposed mechanical option. Some significant issues are discussed below [68].

8.3.1. Cost

Cost of another innovation is constantly high at first and individuals require some an opportunity to acknowledge the things. The cost will be let down if its utilization increments and that will when everybody thinks about its favorable position and employments. The high-volume producing cuts cost down the main answer for this issue, additionally one needs starting applications without dug in minimal effort options. Innovation utilized as a part of it is mind-boggling and in addition it needs a great deal of cost for upkeep. Energy components themselves might be just 25% of aggregate framework cost.

8.3.2. Fuels for Fuel Cells

The greater part of the power devices utilizes just unadulterated hydrogen as a fuel for working and this fuel isn't accessible in vast amounts as will be required for the best possible working and for all the satisfaction of necessities. Typically, it is connected with the "Hydrogen Economy." The other technique is to synthetically change over different energizes to hydrogen yet that will be not efficient as well.

8.3.3. Not Portable

It is extremely hard to store, exchange and transport them from one place to other. A vast space and economy is expected to do as such.

8.4. DIFFERENCE BETWEEN FUEL CELL AND BATTERY

8.4.1. Common Attributes

Photovoltaic power yields change depending predominantly upon sunlight based protection and cell temperature. Since control of the encompassing climate conditions is past people's ability, it is practically unthinkable for human administrators to control the PV control itself. In this way, a PV control generator may now and then experience sharp yield control vacillations attributable to irregular climate conditions, which causes control issues, for example, stack recurrence control, generator voltage control and even framework security investigation. There is, in this way, a requirement for reinforcement control offices in the PV control age. Batteries and energy units are the no doubt advancements to give the PV framework reinforcement control on the grounds that these two reinforcement control sources contain some particular highlights in like manner. Those qualities are recorded beneath [69].

8.4.1.1. Fast Load-Response Capability

Fuel cells and batteries are able to respond very fast to load changes because their electricity is generated by chemical reactions.

8.4.1.2. Modularity in Production

Industrial facility gets together of standard cell units gives energy component and battery control plants with short lead-time from wanting to establishment. This particular creation empowers them to be included discrete additions of limit, which permits better coordinating of the power plant ability to expected load development. Interestingly, the establishment of a solitary extensive traditional power plant may create overabundance limit with regards to quite a long while, particularly if the heap development rate is low.

8.4.1.3. High Reliable Sources

Because of their numerous parallel-secluded units and nonappearance of electromechanical turning masses, energy component and battery control plants are more solid than some other types of energy age. Thus, a utility that introduces various energy component or battery control plants can diminish its hold edge limit while keeping up a steady level of the framework unwavering quality.

8.4.1.4. Site Selection Flexibility

The electrochemical change procedures of energy units and batteries are tranquil on the grounds that they don't have any major pivoting masses. Outer water necessity for their operation is, assuming any, next to no while regular power plants require gigantic measure of water for framework cooling. Along these lines, they can lessen or take out water quality issues made by the customary plants' warm releases. Air poison outflow levels of power modules and batteries are none or practically nothing. Discharges of 2 SO and X NO in the energy component control plant are values are anticipated to be around 1,000 times littler than those of non-renewable energy source control plants since power devices don't depend on a fuel-copying process. These earth kindhearted attributes make it workable for those power plants to be found near load focuses in urban and rural zone. It can likewise decrease vitality misfortunes and expenses related with transmission and circulation hardware. These sitting close load focuses may likewise lessen the probability of framework power outages.

8.4.2. Different Attributes

Electric current is created in a capacity battery by synthetic responses. A similar substance responses happen in an energy component, yet there is a contrast between them regarding fuel stockpiling. Away batteries compound vitality is put away in the positive/negative cathodes of the batteries. In power modules, in any case, the fills are put away outside the cells and should be sustained into the terminals constantly when the energy units are required to produce power. Other itemized correlation between battery reinforcement and energy component reinforcement for PV control supplement is made in the accompanying segments.

8.4.2.1. Efficiency

Power age in energy components straightforwardly change over accessible synthetic free vitality to electrical vitality as opposed to experiencing heat trade forms. Subsequently, one might say that energy components are a more proficient power transformation innovation than the ordinary steam-applying power ages. Figure 8.1 represents vitality transformation forms for a traditional power generator and an energy unit. While the energy unit is a one-advance procedure to create power, the ordinary power generator has a few stages for power age and each progression requires a specific measure of vitality misfortune. Energy unit control frameworks have around 40–60% efficiencies relying upon the sort of electrolytes and free of size.

Battery control frameworks themselves have high vitality efficiencies, yet their general framework efficiencies from crude fuel (for the most part coal or atomic) through the batteries to change over air conditioning power are diminished to beneath 30%. This is on the grounds that vitality misfortunes happen at whatever point one vitality shape is changed over to another.

8.4.2.2. Capacity Variation

As the battery releases, its terminal voltage, continuously diminishes. The fall of the terminal voltage on release is because of its inside protection. Be that as it may, the interior protection of a battery changes with its cell temperature and condition of release. The lessening in battery voltages with expanding release streams and furthermore a decrease in battery limit with expanding rate of release is very extreme and can be plainly observed. For energy component control frameworks, they have similarly high productivity at both fractional and full loads. The client's interest for electrical vitality isn't generally steady. Along these lines, for a power utility to keep acclimation to this evolving request, either huge base-stack control plants should some of the timework at part stack, or littler cresting units must be utilized amid times of popularity. Both way, proficiency endures and contamination increments. Energy component frameworks have a more prominent proficiency at full load and this high productivity is held as load lessens, so wasteful topping generators may not be required.

8.4.2.3. Operation Flexibility

Power modules have preference over capacity batteries in the regard of operational adaptability. Batteries require a few hours to be taken for reviving after they are completely released. Amid release the batteries' cathode materials are lost to the electrolyte, and the terminal materials can be recouped amid the energizing procedure. Power devices, then again, don't experience such material changes. The fuel put away outside the cells can rapidly be renewed, so they don't rundown as long as the fuel can be provided. The power devices demonstrate higher vitality thickness than the batteries when they work for over 2 hours. It implies that power device control frameworks with generally little weight and volume can create expansive vitality yields. It is to stifle the PV control vacillations because of the progressions of sun oriented force and cell temperature. The way that the PV control yields change strongly under harsh climate conditions makes it difficult to choose the limit of the battery control plants since their releasing rates are not steady. For a lead-corrosive battery, the most appropriate battery innovation for photovoltaic applications to date, the profundity of release ought not surpass 80% on the grounds that the profound release cycle diminishes its compelling lifetime. To keep the profound release and to supplement fluctuating the PV powers created on severe climate days, the battery limit must be huge. The two diverse PV control varieties, the specked bend requires a bigger battery limit than the other bend. Besides, the vast battery limit is typically not completely used, but rather for just a few days. Energy units coordinated with photovoltaic frameworks can give smoother operation. The energy component framework is fit for reacting rapidly enough to level the joined power yield of the half and half PV-power device framework if there should be an occurrence of extreme changes in PV control yield.

8.4.2.4. Cost Difference

The historical backdrop of power module gear costs has demonstrated that the cost of energy units has dropped altogether as the business showcase develops and the assembling innovation winds up plainly develop. Starting expense of phosphoric corrosive energy unit control plants was \$5,500/kW and the present framework cost is about \$3,000/kW [8]. This cost is relied upon to diminish further to around \$1,500/kW in future [9]. A lead-corrosive battery control plant has as of now the least battery cost at around \$150/kWh on the grounds that it has been the longest and most completely created battery innovation. The battery cost is anticipated to decrease to \$100/kWh later on. Nickel-cadmium (NiCd) batteries are 4 to 5 times more costly than the lead-corrosive sorts. Once the NiCd batteries are completely develop, their cost will drop however they won't be as low as the lead-corrosive ones on account of the crude material cost.

CHAPTER 9

ELECTROCHEMICAL ASPECTS OF FUEL CELLS

CONTENTS

9.1. Energy And Heat Balance System 90
9.2. Change In Gibb's Free Energy 92
9.3. Principles Of Thermodynamics Related To Fuel Cells 95
9.4. Fuel Cell Reactions 98

In this chapter, the fundamental thermodynamic and electrochemical standards behind power module operation and innovation are portrayed. The essential electrochemistry standards deciding the operation of the power device, the energy of redox responses amid the energy component operation, the mass and vitality transport in a power module, and so forth., are portrayed quickly to give a comprehension of useful energy component frameworks. The perfect and useful operation of power modules and their productivity are additionally portrayed. This will give the structure to comprehend the electrochemical and thermodynamic nuts and bolts of the operation of power devices and how energy component execution can be affected by the working conditions. The impact of thermodynamic factors like weight, temperature, and gas focus, and so forth., on energy unit execution must be dissected and comprehended to anticipate how power modules interface with the frameworks where it is connected. Understanding the effect of these factors permits framework examination investigations of a particular power module application.

9.1. ENERGY AND HEAT BALANCE SYSTEM

All power age frameworks require a vitality adjust to show the working of the framework in detail. In a comparative mold, the power device framework requires a vitality or warmth adjust investigation. The vitality adjust examination in the power device ought to be founded on vitality change forms like power age, electrochemical responses, warm misfortune, and so forth. The vitality adjusts investigation changes for the diverse sorts of power modules on the grounds that the different sorts of electrochemical responses happen as per the energy component sort. The enthalpy of the reactants entering the framework should coordinate the entirety of the enthalpies of the items leaving the phone, the net warmth produced inside the framework, the dc control yield from the phone, and the warmth misfortune from the phone to its environment. The vitality adjust examination is finished by deciding the energy unit temperature at the exit by having data of the reactant organization, the temperatures, H_2 and O_2 usage, the power created, and the warmth misfortune. The energy component response reverse of the electrolysis response is a compound procedure that can be partitioned into two electrochemical half-cell responses. The most straightforward and regular response experienced in energy components is

$$H_2 + \frac{1}{2}O_2 \rightarrow H_2O \tag{1}$$

Examining from a thermodynamic perspective, the greatest work yield got from the above response is identified with the free-vitality change of the response. Treating this examination regarding the Gibbs free vitality is more valuable than that as far as the adjustment in Helmholtz free vitality, since it is more functional to do concoction responses at a consistent temperature and weight instead of at steady temperature and volume. The above response is unconstrained and thermodynamically supported on the grounds that the free vitality of the items is not as much as that of the reactants. The standard free vitality change of the power module response is shown by the equation

$$\Delta G = -nFE \tag{2}$$

where ΔG is the free vitality change, n is the quantity of moles of electrons included, E is the reversible potential, and F is Faraday's consistent. In the event that the reactants and the items are in their standard expresses, the condition can be spoken to as

$$\Delta G^0 = -nFE^0 \tag{3}$$

The estimation of ΔG comparing to Eq. (1) is −229 kJ/mol, n = 2, F = 96500 C/g.mole electron, and thus the computed estimation of E is 1.229 V. The enthalpy change ΔH for an energy component response shows the whole warmth discharged by the response at steady weight. The power device potential as per ΔH is characterized as the thermo-unbiased potential, Et,

$$\Delta H = -nFE_t \tag{4}$$

where E_t has an estimation of 1.48 V for the response spoke to by Eq.(2). The electrochemical responses occurring in an energy component decide the perfect execution of a power device; for various types of powers relying upon the electrochemical responses that happen with various fills, where CO is carbon monoxide, e− is an electron, H_2O is water, CO_2 is carbon dioxide, H+ is a hydrogen particle, O_2 is oxygen, $CO_3^{2-}$is a carbonate particle, H_2 is hydrogen, and OH^- is a hydroxyl particle. It is evident that starting with one sort of cell then onto the next the responses differ, and in this manner do the sorts of fuel. The base temperature for ideal working conditions shifts from cell to cell. This detail will be talked about in consequent sections. Low to medium-temperature energy components, for example, polymer electrolyte fuel cells (PEMFC), antacid power modules (AFC), and phosphoric corrosive power devices (PAFC) are constrained by the necessity of honorable metal electro impetuses for ideal response rates at the anode and cathode, and H_2 is the most prescribed fuel. For high temperature energy components, for example, liquid carbonate power modules (MCFC) and strong oxide energy components (SOFC) the impetus limitations are less stringent, and the fuel sorts can shift. Carbon monoxide can harm a respectable metal electro impetus, for example, platinum (Pt) in low-temperature power devices, however it fills in as a potential fuel in high temperature energy components where non-honorable metal impetuses, for example, nickel (Ni), or oxides can be utilized as impetuses. The perfect execution of an energy component can be spoken to in various ways. The most ordinarily utilized practice is to characterize it by the Nernst potential spoke to as the cell voltage.

The Nernst condition is a portrayal of the connection between the perfect standard potential E_0 for the energy component response and the perfect balance potential E at different temperatures and weights of reactants and items. Once the perfect potential at standard conditions is known, the perfect voltage can be resolved at different temperatures and weights using these conditions. As indicated by the Nernst condition for hydrogen oxidation, the perfect cell potential at a given temperature can be expanded by working the cell at higher reactant weights. Enhancements in power device execution have been seen at higher weights and temperatures. The image E speaks to the balance potential, E_0 the standard potential, P the gas weight, R the general gas consistent, F Faraday's steady and T the supreme temperature.

When all is said in done in a power device the response of H_2 and O_2 produces H_2O. At the point when hydrocarbon fills are engaged with the anode response, CO_2 is additionally created. For liquid carbonate energy units CO_2 is devoured in the cathode response to keep up the invariant carbonate fixation in the electrolyte. Since CO_2 is produced at the anode and expended at the cathode in MCFCs, and in light of the fact that the groupings of the anode and cathode streams are not really equivalent, that is the reason the Nernst condition incorporates the halfway weights of CO_2 for both terminal responses.

9.2. CHANGE IN GIBB'S FREE ENERGY

The perfect standard capability of a H_2 or O_2 power module (E_0) is 1.229 V with fluid water as the item and 1.18 V for water with vaporous item. This esteem is ordinarily alluded to as the oxidation capability of H_2. The potential can likewise be communicated as an adjustment in Gibbs free vitality for the response of hydrogen and oxygen. The adjustment in Gibbs free vitality increments as cell temperature diminishes and the perfect capability of a cell is relative to the adjustment in the standard Gibbs free vitality. It is certain that the impact of temperature on the standard potential is more articulated for high-temperature energy components. This case relates to low, medium, and high-temperature power devices. Consequently the perfect potential is under 1.229 V while considering the vaporous water item in an energy unit. The perfect and real execution of a power device is very extraordinary, particularly when one breaks down the potential current reaction of an energy unit. Electrical vitality is acquired from an energy unit when a current is drawn, however the genuine cell potential is brought down from its harmony potential on account of irreversible misfortunes because of different reasons. A few variables add to the irreversible misfortunes in a down to earth power device. The misfortunes, which are by and large brought polarization or over potential, begin principally from initiation polarization, ohmic polarization, and gas focus polarization (Chase et al., 1985). These misfortunes result in a cell potential for a power device that is not as much as its optimal potential.

9.2.1. Activation Polarization

The first of these three noteworthy polarizations are the enactment misfortune, which is articulated in the low current locale. In this locale, electronic obstructions must be overcome before the approach of present and ionic stream. The initiation misfortune is specifically relative to the expansion in current stream. The actuation polarization can be spoken to as

$$\eta_{act} = \frac{RT}{\alpha nF} \ln\left[\frac{i}{i_0}\right] \quad (5)$$

where η_{act} is the initiation polarization, R the all-inclusive gas steady, T the temperature, α the charge exchange coefficient, n the quantity of electrons included, F the Faraday consistent, I the present thickness, and i_0 the trade current thickness. Initiation polarization is because of the moderate electrochemical responses at the cathode surface, where the species are oxidized or diminished in a power device response. Enactment polarization is specifically identified with the rate at which the fuel or the oxidant is oxidized or lessened. On account of power module responses, the enactment hindrance must be overwhelmed by the responding species [70].

9.2.2. The Ohmic Polarization

The ohmic polarization changes relatively to the expansion in current and increments over the whole scope of streams because of the consistent idea of energy unit protection. The ohmic polarization can be spoken to as

$$\eta_{ohm} = iR_c \tag{6}$$

where η_{ohm} is the ohmic polarization and R_c is the cell protection. The starting point of ohmic polarization originates from the protection from the stream of particles in the electrolyte and stream of electrons through the anodes and the outside electrical circuit. The prevailing ohmic misfortune is in the electrolyte, which is diminished by diminishing the terminal partition, improving the ionic conductivity of the electrolyte and by adjustment of the electrolyte properties (Figure 9.1).

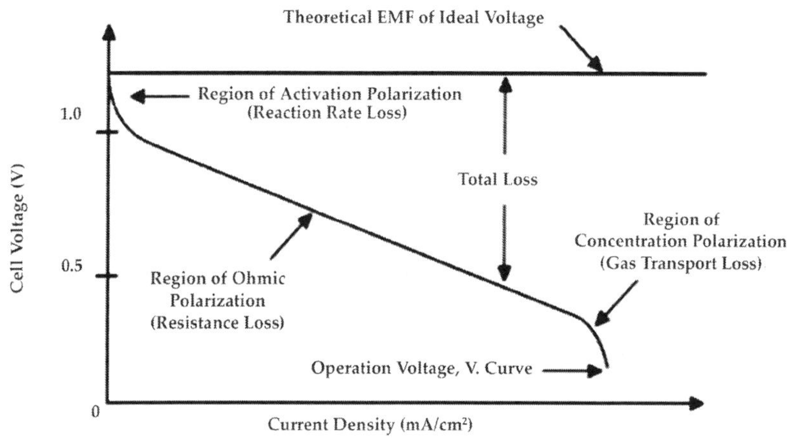

Figure 9.1: Ideal versus actual performance of fuel cell. (Source: Pilatowsky, I. (2014). *Cogeneration Fuel Cell-Sorption Air Conditioning Systems*. [Place of publication not identified]: Springer).

9.2.3. The Concentration Polarization

The fixation misfortunes happen over the whole scope of current thickness, yet these misfortunes end up plainly unmistakable at high restricting streams where it ends up noticeably troublesome for gas reactant stream to achieve the power device response locales. The focus polarization can be spoken to as

$$\eta_{con} = \left[\frac{RT}{nF}\right] ln\left[1 - \frac{i}{i_{Ä}}\right] \quad (7)$$

where η_{con} is the fixation polarization, i_{Δ} is the restricting current thickness. As the reactant gas is devoured at the cathode through the electrochemical response, there will be a potential drop because of the drop in the underlying centralization of the main part of the liquid in the environment. This prompts the development of a focus slope in the framework.

A few procedures are in charge of the arrangement of the fixation polarization. These are

(1) moderate dissemination of the gas stage in the cathode pores.

(2) arrangement of reactants into the electrolyte.

(3) disintegration of items out of the framework.

(4) dissemination of reactants and items, from the response locales, through the electrolyte.

At reasonable current densities, there is ease back transport of reactants to the electrochemical response and moderate expulsion of items from the response site, which is a noteworthy supporter of the fixation polarization. The net aftereffect of focus polarization in current stream in a power module is to expand the anode potential and to diminish the cathode potential. This will bring about the decrease of the phone voltage. The net consequence of fixation polarization in current stream in a power module is to build the anode potential and to diminish the cathode potential. This will bring about the diminishment of the phone voltage [71].

9.2.4. Fuel Crossover Losses

Misfortunes that outcome from the misuse of fuel going through the electrolyte and electron conduction through the electrolyte. This misfortune is normally little, yet can be more imperative in low temperature cells however can be more critical in low temperature cells.

9.3. PRINCIPLES OF THERMODYNAMICS RELATED TO FUEL CELLS

The impact of temperature and weight on the cell potential might be examined on the premise of the Gibbs free vitality variety as for temperature and weight in a power device. This might be composed as

$$[\frac{\partial E}{\partial T}]_P = \frac{\Delta S}{nF} \quad (8)$$

$$[\frac{\partial E}{\partial T}]_T = -\frac{\Delta V}{nF} \quad (9)$$

where −ΔV is the adjustment in volume, ΔS is the entropy change, E is the cell potential, T the temperature, P the reactant gas weight, n the quantity of electrons exchanged, and F Faraday's steady. Since the entropy change for the H_2/O_2 power module response is negative the reversible capability of the H_2/O_2 energy unit diminishes with an expansion in temperature, expecting that the response item is fluid water. For the above response, the volume change is negative, subsequently the reversible potential increments with an expansion in weight. The impact of temperature on the power module voltage is where the energy unit execution information from regular working cells and the reliance of the reversible capability of H_2/O_2 power modules on temperature are given. The cell voltages of PEFC, PAFC, and MCFC demonstrate a solid reliance on temperature. The reversible potential declines with expanding temperature, however the working voltages of these power modules really increment with an expansion in working temperature. PEFC displays a greatest in working voltage. The lower working temperature of SOFC is constrained to around 1000°C since the ohmic protection of the strong electrolyte increments quickly as the temperature diminishes. In any case, progresses in materials science in growing new strong oxide electrolytes and thin film strong electrolytes have prevailing with regards to bringing down the base working temperature of SOFC beneath 1000°C. Regularly power modules work at voltages impressively lower than the reversible cell voltage. The better execution is identified with changes in the sorts of polarizations influencing the cell as the temperature shifts. An expansion in the working temperature is valuable to power device execution due to the expansion in response rate, higher mass exchange rate, and ordinarily bring down cell protection emerging from the higher ionic conductivity of the electrolyte. What's more, the CO resilience of electro catalysts in low-temperature power devices enhances as the working temperature increments.

An expansion in working weight has a few constructive outcomes on power module execution. The halfway weights of reactant gasses, dissolvability, and mass exchange rates are higher at higher weights. The electrolyte misfortune by dissipation is diminished at higher working weights. The framework productivity is expanded by the expansion in weight. The advantages of expanded weight might be contrasted and the issues related with power device materials and other related framework instrumentation. Particularly higher weights increment material issues in MCFC and SOFC. Weight contrasts must be limited to avoid reactant gas spillage through the electrolyte and seals. High weight favors carbon testimony and methane development in the fuel gas. The most extreme electrical work reachable in a power device working at steady temperature and weight is given by the adjustment in the Gibbs free vitality of the electrochemical response [72],

$$W = \Delta G = -nFE \quad (10)$$

where n is the quantity of electrons taking an interest in the response, F is Faraday's consistent (96,487 coulombs/g-mole electron), and E is the perfect capability of the cell. On the off chance that we consider the instance of reactants and items being in the standard state, then

$$\Delta G^0 = -nFE^0 \quad (11)$$

The general responses for this can be utilized to create both electrical vitality and warmth. The most extreme work accessible from a fuel source is identified with the free vitality of response on account of an energy unit, though the enthalpy of response is the applicable amount for a warmth motor, i.e.,

$$\Delta G = \Delta H - T\Delta S \quad (12)$$

where the contrast amongst ΔG and ΔH is relative to the adjustment in entropy ΔS. This entropy change is showed in changes in the degrees of flexibility for the concoction framework being considered. The greatest measure of electrical vitality accessible is ΔG as specified above, and the aggregate warm vitality accessible is ΔH. The measure of warmth that is created by an energy component working reversibly is TΔS. Responses in power devices that have negative entropy change create warm, while those with positive entropy change may separate warmth from their environment [73].

Separating Eq. (12) regarding temperature or weight, and substituting it into Eq. (1) we get

$$[\frac{\partial E}{\partial T}]_P = \frac{\Delta S}{nF} \quad (13)$$

$$[\frac{\partial E}{\partial T}]_T = -\frac{\Delta V}{nF} \quad (14)$$

This was shown earlier too.

9.4. FUEL CELL REACTIONS

Energy unit operation is affected by different thermodynamic and electrochemical factors, for example, temperature, weight, gas fixation, reactant use, current thickness, and so forth., which specifically impact the phone potential and voltage misfortunes. Changing the power module working parameters can have either a useful or an unfavorable effect on energy unit execution and on the execution of other framework parts. Changes in working conditions may bring down the cost of the phone, however increment the cost of the fringe parts [75]. For the most part, a bargain in the working parameters is made to meet the required application. It is conceivable to acquire low framework cost and accomplish satisfactory cell life by working at ideal working conditions. Working conditions are enhanced by characterizing particular framework necessities, for example, control prerequisite level, voltage, current prerequisite and so on. From this and through life cycle thinks about, the power, voltage, and current necessities of the energy component stack and individual cells are resolved. It is an issue of picking an ideal cell working point until the point that the framework necessities are fulfilled. For instance, an outline point at high current thickness will permit a littler cell measure at bring down capital cost to be utilized for the stack, yet a lower framework productivity comes about. This kind of working point would be required by a vehicle application where light weight, little volume, and proficiency are imperative parameters for cost adequacy. Power modules equipped for higher current thickness operation would be of unique intrigue. Operation at a lower current thickness, however higher voltage would be more reasonable for stationary power plant operation. Operation at a higher weight will build cell execution and lower cost.

It is typical and appears to be sensible to outline the cell to work at the most extreme power thickness that crests at a higher current thickness. In any case, operation at the higher power densities will mean operation at bring down cell voltages or lower cell effectiveness. Setting the working point at the pinnacle control thickness may cause insecurity in control in light of the fact that the framework will tend to sway amongst higher and bring down current densities around the pinnacle. It is typical practice to work the cell at a point towards the left half of the power thickness top and at a point that yields a trade off between low working expense and low capital cost [76].

CHAPTER 10

WORKING AND OPERATION OF FUEL CELLS

CONTENTS

10.1. Reversible Cell Potential 100
10.2. Effect Of Temperature On Reversible Cell Potential 105
10.3. Effect Of Pressure On Reversible Cell Potential 108

10.1. REVERSIBLE CELL POTENTIAL

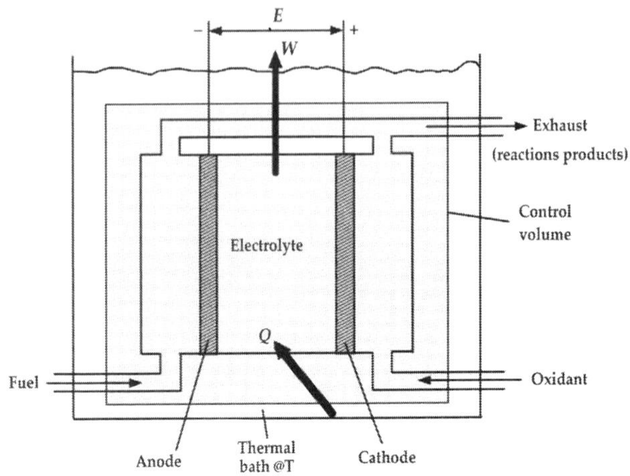

Figure 10.1: Thermodynamic model of fuel cell. (Source: *https://ars.els-cdn.com/content/image/1-s2.0-S1752301X07800068-gr1.jpg*).

In a power device, the substance vitality of a fuel and an oxidant is changed over straightforwardly into electrical vitality, which is shown as far as cell potential and electrical current yield. The most extreme conceivable electrical vitality yield and the relating electrical potential distinction between the cathode and anode are accomplished when the power device is worked under the thermodynamically reversible condition. This greatest conceivable cell potential is called reversible cell potential, one of the altogether critical parameters for power modules [77]. We might apply crucial thermodynamic standards to determine the reversible cell potential in this area. A thermodynamic system demonstrate is appeared in Figure 10.1 for the investigation of power device execution. It is a control-volume framework for the energy unit to which fuel and oxidant streams enter and item or fumes stream exits. The energy unit is situated inside a warm shower with a specific end goal to keep up the coveted framework temperature T. The reactant streams (fuel and oxidant) and the fumes stream are considered to have a similar temperature T and weight P. It is accepted that the fuel and oxidant inflow and the fumes surge are enduring; the motor and gravitational potential vitality changes are unimportant. Further, the general electrochemical responses happening inside the energy component framework limit is depicted as takes after [78].

$$Fuel\left(e.g.,\ H_2\right)+Oxidant\left(e.g.,\ O_2\right)\Delta W+Q+Product \tag{1}$$

where W is the rate of work done by the framework and Q the rate of warmth moved into the framework from the encompassing consistent temperature warm shower, which may, or may not, be in warm harmony with the energy component framework at the temperature T and weight P. For hydrogen/oxygen power devices, the response item is typically water. At that point, the first and second laws of thermodynamics can be composed, separately, for the present power device framework, as

$$\frac{dE_{CV}}{dt}=\left[\left(\dot{N}h+KE+PE\right)_F+\left(\dot{N}h+KE+PE\right)_{Ox}\right]_{in}-\left[\left(\dot{N}h+KE+PE\right)_{Ex}\right]_{out}+\dot{Q}-\dot{W} \tag{2}$$

$$\left[Increase\ in\ System\ energy\right]=\left[Energy\ in\ by\ low\ mass\ flow\right]-\left[Energy\ out\ by\ mass\ flow\right]$$
$$+\left[Energy\ in\ as\ heat\right]+\left[Entropy\ generated\right] \tag{3}$$

$$\frac{dS_{CV}}{dt}=\left[\left(\dot{N}_s\right)_F+\left(\dot{N}_s\right)_{Ox}\right]_{in}-\left[\left(\dot{N}s\right)_{Ex}\right]_{out}+\frac{\dot{Q}}{T}+\wp_s \tag{4}$$

$$\left[Increase\ in\ System\ entropy\right]=\left[Entropy\ in\ by\ mass\ flow\right]-\left[Energy\ out\ by\ mass\ flow\right]+$$
$$\left[Energy\ in\ by\ heat\ transfer\right]+\left[Entropy\ generated\right] \tag{5}$$

where N is the molar stream rate, h (total) enthalpy per unit mole, s the particular entropy on a mole premise and s is the rate of entropy age because of irreversibilities. The subscript "F," "Bull" and "Ex" remain for fuel, oxidant and fumes stream, separately. "KE" and "PE" signify dynamic and gravitational potential vitality that are being completed in and of the framework by the mass stream. For an enduring procedure, there are no worldly changes in the measure of vitality EC.V. also, entropy SC.V. inside the control-volume framework, consequently, $dEC.V./dt_0$ and $dSC.V./dt_0$. Further, the adjustments in the active and gravitational potential vitality are immaterial for the present procedure. In this way, Eqs. (10.2) and (10.3) can be disentangled as takes after [79]

$$\dot{N}_F\left(h_{in}-h_{out}\right)+\dot{Q}-\dot{W}=0 \tag{6}$$

$$\dot{Q}=-T\wp_s-\dot{N}_F\left(s_{in}-s_{out}\right) \tag{7}$$

$$h_{in} = \left[h_F + \frac{\dot{N}_{Ox}}{\dot{N}_F} h_{Ox} \right]_{in} \quad and \quad h_{out} = \frac{\dot{N}_{Fx}}{\dot{N}_F} h_{Ex} \tag{8}$$

where h_{in} is the measure of enthalpy per mole of fuel conveyed into the framework by the reactant inflow and h_{out} is the measure of enthalpy per mole of fuel removed from the framework by the fumes stream. Thus,

$$s_{in} = \left[s_F + \frac{\dot{N}_{Ox}}{\dot{N}_F} s_{Ox} \right]_{in} \quad and \quad s_{out} = \frac{\dot{N}_{Fx}}{\dot{N}_F} s_{Ex} \tag{9}$$

are the measure of entropy per mole of fuel brought into the framework by the reactant in stream, and the measure of entropy per mole of fuel did of the framework by the active fumes stream containing the response items, separately. Substitution of Eq. (10.5) into (10.4) yields

$$\dot{W} = \dot{N}_F \left(h_{in} - h_{out} \right) - \dot{N}_F T \left(s_{in} - s_{out} \right) - T \dot{\wp}_s \tag{10}$$

Now let

$$w = \frac{\dot{W}}{\dot{N}_F}; \; q = \frac{\dot{Q}}{\dot{N}_F}; \; and \; \wp_s = \frac{\dot{\wp}_s}{\dot{N}_F} \tag{11}$$

represent, respectively, the amount of work done, heat transferred and entropy generated per mole of fuel, Eqs. (10.5) and (10.8) then become"

$$q = -T\wp_s - T\left(s_{in} - s_{out} \right) = T\Delta s - T\wp_s \tag{12}$$

$$w = \left(h_{in} - h_{out} \right) - T\left(s_{in} - s_{out} \right) - T\wp_s \tag{13}$$

Because the enthalpy and entropy change for the fuel cell reaction is defined as:

$$\Delta h = h_{out} - h_{in} \quad and \quad \Delta s = s_{out} - s_{in} \tag{14}$$

Eq. (1.11) can also be expressed as

$$w = -\Delta h + T\Delta s - T\wp_s = -\left[\left(h - T_s \right)_{out} - \left(h - T_s \right)_{in} \right] - T\wp_s \tag{15}$$

From the meaning of the Gibbs work (per mole of fuel) g h T_s, Eq. (10.11) or (10.13) can likewise be composed as

$$w = -\left(g_{out} - g_{in} \right) - T\wp_s = -\Delta g - T\wp_s \tag{16}$$

Since by the second law of thermodynamics, entropy can be created, however can never be annihilated, we know $\wp_s = 0$, and additionally the supreme temperature (in Kelvin scale) T 0 by the third law of thermodynamics, the most extreme conceivable work (i.e., valuable vitality) yield from the present framework happens when $\wp_s = 0$, or under the thermodynamically reversible condition, since the adjustment in the Gibbs work is typically negative for helpful power module response. Subsequently, from Eq. (10.14) plainly the most extreme conceivable work yield from the present power device framework is equivalent to the lessening in Gibbs work, or

$$w_{max} = -\Delta g \qquad (17)$$

for every reversible procedure, paying little mind to the particular kind of power devices included. Truth be told, it may be called attention to that in the inference of Eqs. (10.14) and (10.15), no specifics about the control-volume framework have been stipulated; subsequently they are substantial for any vitality transformation frameworks. For a power module framework, the electrical vitality yield is customarily communicated as far as the cell potential contrast between the cathode and the anode [80]. Since (electrical) potential is the (electrical) potential vitality per unit (electrical) charge, its SI unit is J/Coulomb, which is all the more regularly called volt or essentially V. Potential vitality is characterized as the work done when charge is moved starting with one area then onto the next in the electrical field, typically alludes to outside circuits. For the interior circuit of energy units, for example, the one appeared in Figure 10.1, electromotive power is the phrasing frequently utilized, which is likewise characterized as the work done by exchanging one Coulomb positive charge from a low to a high potential. Henceforth, electromotive power likewise has the SI unit of J/Coulomb, or volt. We might embrace the phrasing of cell potential, rather than electromotive power, starting now and into the foreseeable future; and we should utilize the documentation E to speak to the cell potential. Since typically electrons are the particles exchanged that convey electrical charge, we may express the work done by an energy component as follows

$$w\left(\frac{J}{mole\ fuel}\right) = E \times \left(\frac{Coulomb\ of\ electron\ charge\ transferred}{mole\ fuel}\right) \qquad (18)$$

Or

$$w = E \times (n\,N_0 e) = E \times (nF) \qquad (19)$$

where n is the quantity of moles of electrons exchanged per mole of fuel expended, N_0 the Avogadro's number 6.023×10^{23} number of electrons/ mole electron and e the electric charge per electron 1.6021×10^{19} Coulomb/ electron. Since $N_0 e$ 96,487 Coulomb/mole electron F is regularly known as the Faraday consistent, the cell potential progresses toward becoming, from Eq. (10.14) [81]

$$E = \frac{w}{nF} = \frac{-\Delta g - T\wp_s}{nF} \tag{20}$$

Thus, the most extreme conceivable cell potential, or the reversible cell potential E_r, becomes

$$E_r = -\frac{\Delta g}{nF} \tag{21}$$

From the reversible cell potential given above, Eq. (11.17) can also be rewritten as

$$E = E_r - \frac{T\wp_s}{nF} = E_r - \eta \tag{22}$$

Where

$$\eta = \frac{T\wp_s}{nF} \tag{23}$$

is the cell voltage misfortune because of irreversibilities or entropy age. Plainly, the genuine cell potential can be ascertained by subtracting the cell voltage misfortune from the reversible cell potential. Then again, the measure of entropy age per mole fuel devoured can be resolved as

$$\wp_s = \frac{nF\eta}{T} = \frac{nF\left(E_r - E\right)}{T} \tag{24}$$

Hence, the measure of entropy age, speaking to the level of irreversibilities the level of deviation from the glorified reversible condition, for the energy component response process can be measured once the cell potential E and the cell working temperature T are known.

The most vital working conditions that impact energy unit execution are the working temperature, weight, and reactant focuses. Before investigating these impacts, we will initially build up some thermodynamic relations that will be helpful for our examination. Review that Gibbs capacity and enthalpy are characterized as

$$g = h - Ts \quad (25)$$

$$h = u + Pv \quad (26)$$

where u is the inward vitality and v is the particular volume. Consolidating the above conditions together and taking the differential, we get

$$dg = du + Pdv + vdP - Tds - sdT \quad (27)$$

Another major connection in thermodynamics is the Gibbs condition for a basic compressible substance, which is

$$Tds = du + Pdv = dh - vdP \quad (28)$$

Substituting above equations results in

$$dg = vdP - sdT \quad (29)$$

Therefore, we arrive at two important relations for fuel cell analysis

$$\left[\frac{\partial g}{\partial T}\right]_P = -s; \left[\frac{\partial g}{\partial P}\right]_T = v \quad (30)$$

Applying the over two conditions to Gibbs work change for a specific power module response, for example, the summed one up given in Eq. (11.24), we at last obtain

$$\left[\frac{\partial \Delta g}{\partial T}\right]_P = -\Delta s \quad (31)$$

$$\left[\frac{\partial \Delta g}{\partial P}\right]_T = \Delta v \quad (32)$$

where Δs and Δv are the separate changes in the entropy and particular volume between the items and reactants. It may be underscored that the over two relations are acquired without making particular suppositions, for example, perfect gas estimate; consequently they are legitimate for any substances experiencing synthetic responses. They will be used for our following examination of temperature and weight impact on the reversible cell potential [82].

10.2. EFFECT OF TEMPERATURE ON REVERSIBLE CELL POTENTIAL

The reversible cell potential, E_r, is a component of temperature, on the grounds that the adjustment in Gibbs work relies upon the power device working temperature and weight. Hence

$$E_r(T,P) = -\frac{\Delta g(T,P)}{nF} \tag{33}$$

At that point the difference in the reversible cell potential with temperature can be communicated as, fusing

$$\left[\frac{\partial E_r(T,P)}{\partial T}\right]_P = -\frac{1}{nF}\left[\frac{\partial \Delta g(T,P)}{\partial T}\right]_P = \frac{\Delta s(T,P)}{nF} \tag{34}$$

Obviously, the variety of E_r with temperature relies upon the adjustment in entropy for the specific energy unit reaction, and three conceivable circumstances may emerge:

1. On the off chance that s is under 0, like

$$H_2 + \frac{1}{2}O_2 \rightarrow H_2O \tag{35}$$

the reversible cell potential will diminish with cell operation temperature.

2. In the event that s is more noteworthy than 0, then the reversible cell potential will increment with temperature, e.g., for the response

$$C(s) + \frac{1}{2}O_2 \rightarrow CO \tag{36}$$

the entropy change is around 89J/K.

3. On the off chance that s is equivalent to 0, then the reversible cell potential will be autonomous of temperature, similar to the reaction [83]

$$CH_4 + 2O_2 \rightarrow CO_2 + 2H_2O(g) \tag{37}$$

For some valuable electrochemical responses, the entropy change is negative and is practically consistent with the difference in temperature to a decent estimate, gave the temperature change T T_{ref} isn't too vast. At that point might be coordinated from the standard reference temperature, T_{ref} 25°C, to the subjective energy unit working temperature T, while keeping weight P steady,

$$E_r(T,P) = E_r(T_{ref},P) + \left[\frac{\Delta s(T_{ref},P)}{nF}\right](T - T_{ref}) \tag{38}$$

On the other hand, we can extend the reversible cell potential articulation, in Taylor arrangement as far as temperature, T, around the reference temperature, T_{ref}, keeping P constant

$$E_r(T,P) = E_r(T_{ref},P) + \left[\frac{\partial E_r(T_{ref},P)}{\partial T}\right](T - T_{ref}) \tag{39}$$

Entirely, the reversible cell potential at any temperature and weight ought to be resolved from Eq. (10.18) by figuring first the property changes for the specific power module response included. Such a method has been taken after for the hydrogen and oxygen response to shape vaporous water, and the outcomes are displayed in Figure 10.2. Unmistakably, the reversible cell potential to be sure abatements directly as temperature is expanded over a substantial temperature run. In any case, it is seen that the reversible cell potential is bigger for item water as fluid at low temperatures, however it diminishes substantially speedier than the vaporous water as item when temperature is expanded. So that at temperatures marginally above around 373K, the reversible cell potential for fluid water item really ends up noticeably littler. This may appear to be interested, yet it is on the grounds that at such high temperatures pressurization is important to keep the item water in fluid frame as the reactants, hydrogen and oxygen, are bolstered at 1 atm. Additionally see that the basic temperature for water is around 647 K, past which particular fluid state does not exist for water, subsequently the shorter bends for the fluid water case appeared in Figure 10.2. As pointed out before, the entropy change for a large portion of energy component responses is negative; thusly, the reversible cell potential abatements as temperature is expanded. Be that as it may, for nearly couple of responses such as

$$C(s) + \frac{1}{2}O_2(g) \rightarrow CO(g) \tag{40}$$

the entropy change is sure, e.g., s 89J/moleK fuel at the standard reference temperature and weight. Therefore, the reversible cell potential for this sort of responses will increment with temperature. Assume NP and NR speak to the quantity of moles of items and reactants, separately, which are in vaporous state and on a for each mole fuel premise, and N NP NR speaks to the change, per mole fuel, in the quantity of moles of gas species amid the response, at that point as a harsh dependable guideline, it may be expressed that:

1. s is more noteworthy than 0 for N more prominent than 0 (because of expanding issue in light of more atoms in the item), and reversible cell potential increments with temperature.
2. s is under 0 for N under 0 (because of diminishing issue in light of less atoms in the item), and reversible cell potential declines with temperature.
3. s is practically equivalent to 0 for ΔN equivalent to 0, and reversible cell potential is practically autonomous of temperature [84].

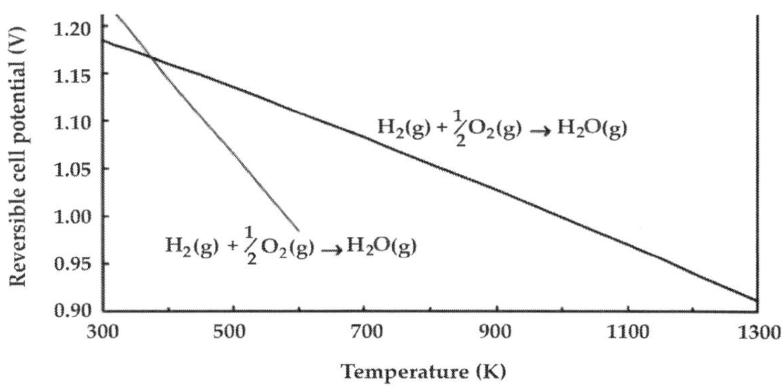

Figure 10.2: Effect of temperature on fuel cell. (Source: *https://www.researchgate.net/profile/Axel_Mertens/publication/224384096/figure/fig1/AS:302782549643289@1449200393774/Fig-1-Comparison-of-the-Efficiency-of-the-Carnot-process-and-the-Fuel-Cell-process.png:*).

10.3. EFFECT OF PRESSURE ON REVERSIBLE CELL POTENTIAL

Taking halfway subsidiary of above equation as for weight while keeping temperature constant, we get

$$\left[\frac{\partial E_r}{\partial P}\right]_T = -\frac{1}{nF}\left[\frac{\partial \Delta g}{\partial P}\right]_T = -\frac{\Delta v}{nF} \tag{41}$$

where

$$\Delta v = v_p - v_g \tag{42}$$

speaks to the volume change of all the vaporous species in the response, on a for every mole fuel premise; the volume of the solids and fluids is considerably littler than the volume of gas species, and can be dismissed for the present reason; v_P and v_R are the particular volume (per mole fuel) of gas items and reactants, individually. Consider all the reactant and item gasses can be dealt with as perfect gasses, can be communicated as

$$\Delta v = v_P - v_R = \frac{N_p \Re T}{p} - \frac{N_R \Re T}{p} = \frac{\Delta N \Re T}{p} \tag{43}$$

where R is the universal constant. Combining Eqs. (10.41) and (10.43) yields

$$\left[\frac{\partial E_r}{\partial P}\right]_T = -\frac{\Delta N \Re T}{nF}\frac{1}{P} \tag{44}$$

Above condition shows that

1. On the off chance that N is more noteworthy than 0, which means item contains a bigger number of moles of gas species than reactant, the reversible cell potential will diminish with weight, for example, for the response

$$C(s) + \frac{1}{2}O_2(g) \rightarrow CO \tag{45}$$

2. In the event that N is under 0, which is the situation for the greater part of helpful power device responses, the reversible cell potential will increment with weight.
3. In the event that N is equivalent to 0, the reversible cell potential won't change with weight.

Note for (1) and (2) over, the measure of progress for reversible cell potential declines bit by bit when weight is expanded. Higher-weight operation brings about mechanical issues, for example, mechanical quality of the phone segments, cell fixing issue, consumption, and so forth. This suggests the execution pick up in E_r at high-weight operation decreases, and may end up plainly unwanted from framework configuration perspective. Presently incorporating Eq. (44) from the standard reference weight, P_{ref} 1 atm, to a self-assertive weight, P, while keeping temperature settled, brings about

$$E_r\left(T,P\right)=E_r\left(T,P_{ref}\right)-\frac{\Delta N\Re T}{nF}\ln\left[\frac{P}{P_{ref}}\right] \tag{46}$$

This demonstrates the weight reliance of the reversible cell potential is a logarithmic capacity, and henceforth the reliance winds up noticeably weaker as weight P is expanded. Further, the weight impact is little on the reversible cell potential at low temperatures, however this impact at high temperatures increments fundamentally on the grounds that the weight impact coefficient, ΔNT/(nF), is straightforwardly corresponding to temperature.

It may be brought up that for high temperature energy components the reliance of the real cell potential Eon the weight takes after nearly the outcomes given in Eq. (45), though a critical deviation happens for the low temperature energy units. The distinction emerges from the way that at high temperatures, the response energy is quick and pressurization essentially expands the reactant fixation, thus better execution straightforwardly. At low temperatures, the response energy is moderate and higher reactant fixation does not yield a relative increment in the cell potential because of the huge cell potential misfortune related with the moderate energy.

CHAPTER 11

EFFICIENCIES AND ENERGY LOSSES

CONTENTS

11.1. Efficiency Of Fuel Cells....112

11.2. Efficiency Of Carnot Cycle....113

11.3. Comparison Of The Efficiencies Of Fuel Cell And Carnot Cycle....115

11.4. Losses Of Fuel Cells....117

11.5. Reversible Energy Loss Compared To Energy Efficiency....118

11.6. Irreversible Energy Loss Compared To Energy Efficiency....119

11.1. EFFICIENCY OF FUEL CELLS

The warm productivity of a vitality transformation gadget is characterized as the measure of helpful vitality created in respect to the change in put away substance vitality, usually alluded to as warm vitality, that is discharged when a fuel is responded with an oxidant. Subsequently the effectiveness might be characterized as

$$\eta_e = \frac{useful\ output\ energy}{\Delta H} \quad (1)$$

Hydrogen (fuel) and oxygen (oxidant) can exist in each other's quality at room temperature, however, in the event that warmed to over 500°C and at a high weight, they detonate fiercely. The ignition response for these gasses can be compelled to happen underneath 500°C within the sight of a fire, for example, in a warmth motor. On account of an energy unit, an impetus can expand the rate of response of H_2 and O_2 at temperatures lower than 500°C in the surrounding of an electrolyte. In high-temperature energy components, a noncombustible response can happen at temperatures more than 500°C in light of controlled detachment of the fuel and oxidant. The procedure occurring in a warmth motor is warm, whereas the power device process is electrochemical. The distinction in these two procedures in vitality transformation is the reality behind effectiveness correlation for these two frameworks. In the perfect instance of an electrochemical vitality transformation response, for example, a power module the adjustment in Gibbs free vitality of the response is accessible as valuable electric vitality at the yield of the gadget. The perfect proficiency of an energy component working irreversibly might be expressed as

$$\eta_e = \frac{\Delta G}{\Delta H} \quad (2)$$

The most commonly used way of expressing efficiency of a fuel cell is based on the change in the standard free energy for the cell reaction

$$H_2 + 1/2O_2 \rightarrow H_2O \quad (3)$$

$$\Delta G^0 = G^0_{H_2O} - G^0_{H_2} - \frac{1}{2}G^0_{O_2} \quad (4)$$

where the item water is in fluid frame. At standard states of response, the compound vitality in the hydrogen/oxygen response is 285.8 kJ/mole and the free vitality accessible for helpful work is 237.1 kJ/mole. Subsequently, the warm effectiveness of a perfect energy component working reversibly on unadulterated hydrogen and oxygen at standard conditions would be

$$\eta_e = \frac{237.1}{285.8} = 0.83 \quad (5)$$

The productivity of a real energy unit can be communicated regarding the proportion of the working cell voltage to the perfect cell voltage. The genuine cell voltage is not as much as the perfect cell voltage due to the misfortunes related with cell polarization and the iR misfortune, as talked about in the before area. The warm proficiency of the energy component would then be able to be composed as far as the genuine cell voltage

$$\eta_e = \frac{useful\ output\ energy}{\Delta H} = \frac{useful\ output\ energy}{\frac{\Delta g}{0.83}} = \frac{V_{cell}.I}{\frac{V_{ideal}.I}{0.83}} = \frac{V_{cell}.0.83}{V_{ideal}} \quad (6)$$

As specified before, the perfect voltage of an energy component working reversibly with unadulterated hydrogen and oxygen in standard conditions is 1.229 V. Subsequently, the warm proficiency of a real power module working at a voltage of V_{cell}, in light of the higher warming estimation of hydrogen is given by

$$\eta_e = \frac{0.83.V_{cell}}{V_{ideal}} = \frac{0.83.V_{cell}}{1.229} = 0.675.V_{cell} \quad (7)$$

An energy unit can be worked at various current densities; the relating cell voltage at that point decides the power module effectiveness. Diminishing the present thickness builds the cell voltage, in this way expanding the power device effectiveness. Truth be told, as the present thickness is diminished, the dynamic cell territory must be expanded to acquire the coveted measure of energy [85].

11.2. EFFICIENCY OF CARNOT CYCLE

The Carnot cycle is a hypothetical thermodynamic cycle. It is the most effective cycle for changing over a given measure of warm vitality into work, or on the other hand, making a temperature contrast, e.g., refrigeration, by doing a given measure of work (Figure 11.1). A Carnot cycle going about as a warmth motor. The cycle happens between a hot repository at temperature T_H and a chilly supply at temperature T_C [86].

It isn't a down to earth motor cycle on the grounds that the warmth move into the motor in the isothermal procedure is too ease back to be of functional esteem. As Schroeder puts it "So don't try introducing a Carnot motor in your auto; while it would expand your gas mileage, you would be passed on the roadway by people on foot."

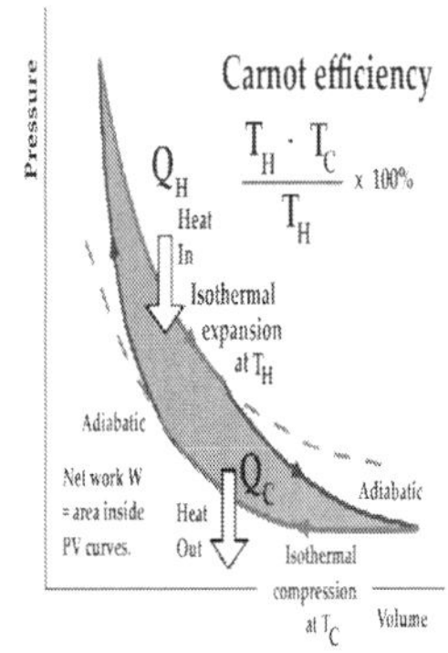

Figure 11.1: Efficiency of Carnot cycle. (Source: *http://hyperphysics.phy-astr.gsu.edu/hbase/thermo/imgheat/carnot2.gif*).

The Carnot cycle is a theoretical, romanticized warm motor that has the most extreme conceivable proficiency [87].

So as to boost productivity, we need to stay away from irreversible procedures, for example, warm exchange with a temperature change.

In this manner, each warmth exchange must be isothermal at either T_H or T_C. The Carnot cycle has two isothermal and two adiabatic procedures, both reversible.

With a specific end goal to figure the effectiveness, we have to discover the proportion Q_C/Q_H.

1. Isothermal extension stomach muscle: warm QH provided from hot supply at steady temperature T_H.

$$Q_H = W_{ab} = nRT_H \ln\frac{V_b}{V_a} \tag{8}$$

2. Isothermal compression W_{cd} is a warm Q_C rejected to icy repository at consistent temperature T_C.

$$Q_C = W_{cd} = nRT_C \ln\frac{V_c}{V_d} \tag{9}$$

So,

$$\frac{Q_C}{Q_H} = -\frac{T_C}{T_H}\frac{\ln\frac{V_c}{V_d}}{\ln\frac{V_b}{V_a}} \tag{10}$$

Now, for the two adiabatic processes $Q = 0$ and 1. Adiabatic expansion b_c

$$T_H V_b^{\gamma-1} = T_C V_c^{\gamma-1} \tag{11}$$

3. Adiabatic compression is

$$T_H V_a^{\gamma-1} = T_C V_d^{\gamma-1} \tag{12}$$

So,

$$\left(\frac{V_b}{V_a}\right)^{\gamma-1} = \left(\frac{V_c}{V_d}\right)^{\gamma-1} \quad or \quad \frac{V_b}{V_a} = \frac{V_c}{V_d} \tag{13}$$

Hence the two logarithms cancel out, and we get

$$\frac{Q_C}{Q_H} = -\frac{T_C}{T_H} \quad or \quad \frac{|Q_C|}{|Q_H|} = -\frac{T_C}{T_H} \tag{14}$$

So the efficiency is

$$e = 1 + \frac{Q_C}{Q_H} = 1 - \frac{T_C}{T_H} = \frac{T_H - T_C}{T_H} \tag{15}$$

The effectiveness of a Carnot motor depends just on the temperature contrast of the two warmth supplies. Along these lines, we can infer that The greater the temperature distinction, the more prominent the proficiency [88].

11.3. COMPARISON OF THE EFFICIENCIES OF FUEL CELL AND CARNOT CYCLE

As natural concerns get expanding consideration, the requirement for growing new innovations that address the conflicting issues of vitality generation and security of the earth winds up noticeably obvious. The remarkable natural quality and high efficiency of power devices make them a potential option vitality hotspot for both stationary and transportation applications. A power module is an electrochemical gadget that changes over the compound vitality of a response straightforwardly into electrical vitality through a controlled concoction response.

A usually expressed preferred standpoint of power modules is that they are more efficient than a comparable procedure of combusting the fuel and changing over the warmth into power. Frequently, this preferred standpoint is clarified by saying 'energy components are not constrained by the Carnot efficiency,' once in a while with the additional remark, 'in light of the fact that a power device isn't a warmth motor.' While the last is valid, the previous is a myth. Energy units and warmth motors are both compelled by a similar most extreme efficiency. The farthest point is set up by the second law of thermodynamics, and neither one of the processes can infringe upon this law.

We acknowledge the way that the power module and the Carnot warm motor efficiencies are both obliged by the second law of thermodynamics, and neither one of the ones can overstep this law. Be that as it may, this announcement does not mean the two frameworks ought to have a similar greatest warm efficiency while being sustained similar measures of substance reactants. Henceforth, the above proclamation isn't a myth. The inborn distinction between the power devices (electrochemical frameworks) and warmth motors, ignition motors, efficiencies is an essential one with respect to the transformation of compound vitality of responses into electrical work. The sole purpose behind this is because of the ignition irreversibility of the last. A time of postponements in using the electrochemical vitality change were because of innovative difficulties. Not perceiving its potential, this offered ascend to an utilization of more than double the fuel that would have been utilized electrochemically to create the same electrical vitality; a large number of the negative results of our present innovation, for example, brown haze, smoke, vibration, and clamor could have been maintained a strategic distance from. We audit and remark on the investigation displayed by Lutz et al. Their determination depends on a perfect ignition process; they have likewise neglected to consider burning irreversibility and the impact of concoction balance criteria, which impelled them to achieve a wrong conclusion. They have disregarded the way that a burning procedure, regardless of whether finish or in balance condition, couldn't be made reversible by supplanting a high-temperature store with an ignition reactor as delineated by them, as contrasted and the first warmth motor, in light of the fact that the burning procedure is innately irreversible. A first-law calculation in light of the perfect burning condition is regularly very deceptive, since the real items as often as possible don't coordinate those anticipated by the hypothetical response. A right hypothetical forecast of conceivable conduct can be accomplished using the second law of thermodynamics; concoction harmony criteria are of significant significance in the investigation of procedures that included compound responses [89].

In the substance response completed in an ignition motor, there is warm exchange, where in a power device, there is charge exchange. In control exchange, the accessible vitality of the electrons can be spent; that is, the electrons fall through the aggregate potential contrast, as on account of the potential vitality of a falling article gave that the inward protection of the energy component and grinding on the falling item are immaterial. Because of the potential distinction between the two terminals in the power device, the additional vitality of the electrons is totally spent when they get around the work circuit. In actuality, since the temperature of the earth is fixed and isn't zero, the efficiency of a warmth motor is constantly less then solidarity.

11.4. LOSSES OF FUEL CELLS

The losses or the misfortunes in an energy component can be partitioned into fuel hybrid and inner streams, initiation misfortunes, ohmic misfortunes and mass transport misfortunes. Fuel hybrid and inside current misfortunes result from the stream of fuel and electric current in the electrolyte. The electrolyte should just transport particles, however a specific fuel and electron stream will dependably happen. In spite of the fact that the fuel misfortune and inside streams are little, they are the primary purpose behind the genuine open circuit voltage (OCV) being lower than the hypothetical one (E_{rev}).

Enactment misfortunes are caused by the gradualness of the responses occurring on the cathode surface. The voltage diminishes to some degree because of the electrochemical response energy. This can be found in the left-hand segment of the present voltage bend above.

The ohmic misfortunes result from protection from the stream of particles in the electrolyte and electrons through the cell equipment and different interconnections. The relating voltage drop is basically corresponding to current thickness, henceforth the expression "ohmic misfortunes."

Mass transport misfortunes result from the decline in reactant focus at the surface of the terminals as fuel is utilized. At greatest (restricting) current, the focus at the impetus surface is for all intents and purposes zero, as the reactants are expended when they are provided to the surface.

Vitality misfortune in power modules happens under both reversible and irreversible conditions. We will examine each kind of vitality misfortune components and related articulation for vitality transformation effectiveness in energy components. Since this part is committed to thermodynamic examination, the reversible misfortune component has been and will be exhibited in subtle elements, while the irreversible misfortune instruments will be portrayed quickly just in this section.

11.5. REVERSIBLE ENERGY LOSS COMPARED TO ENERGY EFFICIENCY

The vitality misfortune in power modules under reversible condition is equivalent to the warmth exchanged (or lost) to the earth, in view of the negative entropy change for the energy unit response. The related vitality transformation proficiency, which has been known as the reversible effectiveness, has been determined as

$$\eta_r = \frac{\Delta g}{\Delta h} = \frac{\Delta g}{\Delta g + T\Delta s} \quad (16)$$

Dividing the numerator and the denominator by the factor "nF," and utilizing above equations we get

$$\eta_r = \frac{E_r}{E_r - T\left(\frac{\partial E_r}{\partial T}\right)_P} \quad (17)$$

Along these lines, when the entropy change is negative, as portrayed prior, the reversible productivity, ηr, is under 100% and the reversible cell potential abatements with temperature; and as per Eq. (17), the reversible proficiency, η_r, likewise diminishes with temperature. For instance, for H2 and O2 response shaping vaporous water at 1atm pressure

$$\left(\frac{\partial E_r}{\partial T}\right)_P = -0.2302 \times 10^{-3} V / K \quad (18)$$

at 25°C, and the reversible effectiveness is around 95% at 25°C, and it ends up plainly 88% at 600K and 78% at 1000K.

For the response of carbon and oxygen to frame carbon monoxide, as appeared in equation above, the entropy change is sure, and the reversible cell potential increments with temperature, as displayed already; subsequently, the reversible effectiveness will likewise increment with temperature. Be that as it may, as examined beforehand, the productivity articulation, is not by any means legitimate for such responses. From the reversible vitality proficiency, and partitioning the numerator and denominator by the factor nF, we have, in the wake of using the reversible cell potential.

$$\eta_r = \frac{E_r}{-\frac{\Delta h}{nF}} = \frac{E_r}{E_{tn}} \qquad (19)$$

where

$$E_{tn} = -\frac{\Delta h}{nF} \qquad (20)$$

is called thermoneutral voltage (or potential), a voltage a power device would have if all the compound vitality of the fuel and oxidant is changed over to electric vitality (i.e., 100% vitality transformation into power).

It ought to be accentuated that from the previous investigation it is realized that for most power module responses, the reversible productivity, ηr, diminishes as the energy unit working temperature is expanded. This impact is vital in considering high temperature energy units, specifically the liquid carbonate power modules (MCFCs) and the strong oxide energy components (SOFCs). For instance, the reversible cell proficiency is lessened to bring down 70% (in view of LHV) for hydrogen and oxygen response at the run of the mill working temperature of 1000°C for SOFCs, instead of around 95% at 25°C as talked about above. This noteworthy decrease in the reversible cell proficiency appears against high temperature power modules. In any case, the irreversible misfortunes, to be portrayed underneath, diminish definitely as temperature is expanded, so the down to earth energy component execution, (for example, proficiency and influence yield under commonsense working condition) increments. Accordingly, promote examination ought to be required for proficiency under down to earth working condition as opposed to the romanticized reversible condition, which is the concentration of the accompanying discussion.

11.6. IRREVERSIBLE ENERGY LOSS COMPARED TO ENERGY EFFICIENCY

For power devices, the reversible cell potential and the comparing reversible effectiveness are gotten under the thermodynamically reversible condition, suggesting that there is no thorough event of ceaseless response or electrical current yield. For useful applications, a helpful measure of work (electrical vitality) is acquired just when a sensibly huge current I is drawn from the cells in light of the fact that the electrical vitality yield is through the electrical power yield, which is characterized as Power EI or Power thickness EJ However, both the cell potential and productivity diminish from its relating (balance) reversible esteems due to irreversible misfortunes when current is expanded. These irreversible misfortunes are frequently called polarization, overpotential or overvoltage in writing, and they begin basically from three sources: enactment polarization, ohmic polarization and focus polarization. The real cell potential as a component of current is the consequences of these polarizations; hence, a plot of the phone potential versus current yield is customarily called a polarization bend. It ought to be seen that the greatness of electrical current yield depends to a great extent on the dynamic cell zone, along these lines, a superior measure is the present thickness, J (A/cm^2), instead of current, I, itself; and the unit A/cm^2 is frequently utilized instead of A/m^2 as the unit for the present thickness since square meter is too substantial to be in any way utilized for power module examination. The perfect cell potential-current connection is free of the current drawn from the cell, and the cell potential stays equivalent to the reversible cell potential. The contrast between the thermoneutral voltage and the reversible cell potential speaks to the vitality misfortune under the reversible condition (the reversible misfortune). Be that as it may, the genuine cell potential is littler than the reversible cell potential and reductions as the current attracted is expanded because of the three components of irreversible misfortunes: actuation polarization, ohmic polarization and fixation polarization. The initiation polarization, η_{act}, emerges from the moderate rate of electrochemical responses, and a segment of the vitality is lost (or spent) on driving up the rate of electrochemical responses keeping in mind the end goal to meet the rate required by the present request. The ohmic polarization, η_{ohm}, emerges because of electrical protection in the cell, including ionic protection from the stream of particles in the electrolyte and electronic protection from the stream of electrons in whatever remains of the cell segments. Regularly, the ohmic polarization is directly reliant on the cell current. Fixation polarization, η_{conc}, is caused by the moderate rate of mass move bringing about the exhaustion of reactants in the region of dynamic response locales and the over accumulation of response items which hinder the reactants from achieving the response destinations. It generally winds up noticeably noteworthy, or even restrictive, at high current thickness when the moderate rate of mass exchange can't take care of the appeal required by the high current yield. As appeared in Figure 11.2, fixation polarization is regularly the reason for cell potential diminishing quickly to zero. The present (thickness) relating to the zero-cell potential is regularly called the restricting current (density), and clearly it is controlled by the focus initiation. From Figure 11.2, it is additionally evident that enactment polarization happens at little current thickness, while focus polarization happens at high current thickness. The straight drop in the cell potential because of protection misfortune happens at prompt current thickness, and functional energy unit operation is quite often situated inside the ohmic polarization locale.

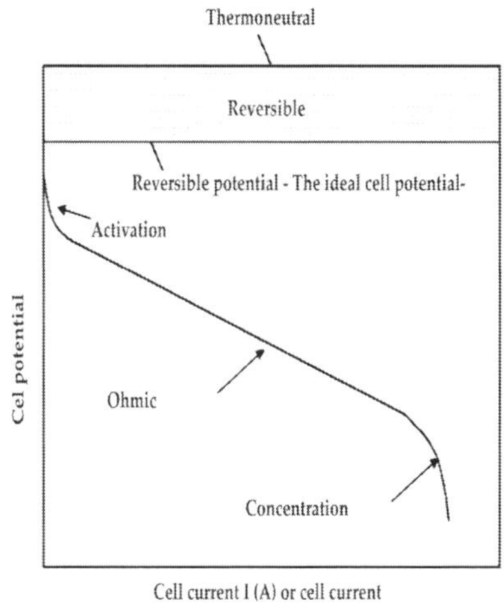

Figure 11.2: Polarization curve. (Source: *https://www.researchgate.net/profile/M_Hashem_Nehrir/publication/3270710/figure/fig2/AS:394697483800584@1471114621638/Fig-2-Polarization-curve-for-a-PEM-hydrogen-oxygen-fuel-cell.pbm*)

Figure 11.2 likewise shows that even at zero current yield from the energy component, the real cell potential is littler than the admired reversible cell potential. This little distinction in cell potential is specifically identified with the substance potential contrast between the cathode and anode. So that even at zero outer load current, there are electrons conveyed to the cathode, where oxygen particles are framed, and relocate through the electrolyte to the anode where they deionize to discharge electrons. The electron discharged moves back to the cathode to proceed with the procedure or "trade." The ionization/deionization responses going before at a moderate rate yield a to a great degree little present, regularly called trade current I_0 or trade current thickness J_0, and the phone potential is diminished beneath the reversible cell potential. Along these lines, trade current emerges from the way that electrons relocating through the electrolyte instead of through the outside load, and around 0.–0.2V of cell potential misfortune comes about because of the trade procedure. Thus, the productivity of a genuine power device is around (8–16%) lower than the reversible cell effectiveness, η_r, even at near zero current yield. The trade current thickness J_0 is little; it is at any rate around 102 A/cm^2 for H_2 oxidation at the anode, and around 105 times slower for O_2 lessening at the cathode. In examination, the O_2 lessening process at the cathode is slow to the point that contending anodic responses assume a huge part, for example, oxidation of electro impetus, consumption of terminal materials, oxidation of natural polluting influences in the anode structure. All these anodic responses result in the erosion of cathodes, in this way restricting the cell life unless fitting counter-measures are taken. It ought to be called attention to that the phone potential misfortune coming about because of the trade current reduces when the current drawn through the outside load is expanded past a specific basic esteem. As the outside current is expanded, the cell potential declines as appeared in Figure 11.2, thus the main impetus for the trade current is diminished, prompting a littler trade current – this is the main type of vitality misfortunes that reductions when outer current is expanded.

From the above talk, it turns out to be evident that the genuine cell potential E is lower than the reversible cell potential E_r, and the distinction is because of the potential misfortunes because of the above irreversible misfortune instruments.

CHAPTER 12

TYPES OF FUEL CELLS

CONTENTS

12.1. Solid Oxide Fuel Cell (SOFC) 124
12.2. Phosphoric Acid Fuel Cell (PAFC) 129
12.3. Polymer Electrolyte Membrane Fuel Cell (PEMFC) 135
12.4. Alkaline Membrane Fuel Cell (AFC) 137
12.5. Molten Carbonate Fuel Cell (MCFC) 144

12.1. SOLID OXIDE FUEL CELL (SOFC)

Strong oxide power devices (SOFCs) offer a spotless, low-contamination innovation to electrochemically produce power at high efficiencies; since their efficiencies are not constrained by the Carnot cycle of a warmth engine. These energy units give many favorable circumstances over conventional vitality transformation frameworks including high effectiveness, dependability, particularity, fuel flexibility, and low levels of NOx and SOx discharges. Calm, without vibration operation of SOFCs likewise takes out commotion for the most part connected with ordinary power age frameworks. Up until around six years prior, SOFCs were being produced for operation principally in the temperature scope of 900 to 1000°C; notwithstanding the capacity of inside changing hydrocarbon energizes (e.g., petroleum gas), such high temperature SOFCs give top notch fumes warmth to cogeneration, and when pressurized, can be coordinated with a gas turbine to additionally expand the general proficiency of the power framework. In any case, decrease of the SOFC working temperature by at least 200°C permits utilization of a more extensive arrangement of materials, is less-requesting on the seals and the adjust of-plant segments, disentangles warm administration, helps in quicker start up and chill off, and brings about less corruption of cell and stack segments. In light of these points of interest, action in the improvement of SOFCs fit for working in the temperature scope of 650 to 800°C has expanded drastically over the most recent couple of years. Be that as it may, at bring down temperatures, electrolyte conductivity and terminal energy diminish altogether; to beat these disadvantages, elective cell materials and plans are by and large broadly researched. A SOFC basically comprises of two permeable cathodes isolated by a thick, oxide particle leading electrolyte. Oxygen provided at the cathode or the terminal of air responds with approaching electrons from the outside circuit to frame oxide particles, which move to the anode or the terminal of a fuel through the oxide particle-leading electrolyte. At the anode, oxide particles consolidate with H_2 or potentially CO in the fuel to shape H_2O as well as CO_2, freeing electrons. Electrons (power) spill out of the anode through the outside circuit to the cathode. The materials for the cell segments are chosen in view of reasonable electrical leading properties expected of these parts to play out their planned cell capacities; satisfactory substance and basic strength at high temperatures experienced amid cell operation and additionally amid cell manufacture; negligible reactivity and inter diffusion among various segments; and coordinating warm development among various parts [90].

12.1.1. Construction

Yttria doped zirconia (YSZ) remains the most broadly utilized material for the electrolyte in SOFCs in view of its adequate ionic conductivity, synthetic steadiness, and mechanical quality. The main downside of settled ZrO_2 is the low ionic conductivity in the lower cell operation temperature administration underneath around 750°C. Two arrangements that have been attempted to determine this issue are to diminish the thickness of the YSZ electrolyte and to discover different acceptors to supplant Y. Scandia-doped zirconia has higher conductivity than YSZ however high cost of scandium and unfavorable maturing impacts in Scandia doped ZrO_2 make it less alluring in commercializing SOFCs. Gadolinium-or samarium-doped CeO_2 materials have higher oxide particle conductivity, e.g., Ce0.9 Gd0.1 O1.95: 0.025 $\Omega^{-1}cm^{-1}$ at 600°C contrasted with zirconia based materials that is 0.005 $\Omega^{-1}cm^{-1}$. Notwithstanding, CeO_2 based materials, under lessening conditions at high temperatures, display critical electronic conductivity and dimensional change because of the arrangement of oxygen opening and the related response of Ce^{4+} to Ce^{3+}. Operation at temperatures beneath around 600°C conquers this issue, and ceria based materials are effectively being utilized as electrolyte in SOFCs by Ceres Power Limited (UK). Notwithstanding the generally utilized fluorite structure oxides, for example, zirconia and ceria, perovskite structure (ABO_3) additionally offers a chance to create oxide particle leading electrolytes by specifically substituting either the A or B particle by isovalent or aliovalent cations. (La, Sr) (Mg, Ga) O_3 (LSMG) has pulled in consideration as an electrolyte since its revelation. LSMG electrolytes, be that as it may, have two disadvantages: questionable cost of gallium, and indeterminate compound and mechanical solidness of LSMG. Regardless of these disadvantages, Mitsubishi Materials Corporation (Japan) is utilizing LSMG as the electrolyte in its SOFCs and has effectively developed and tried to 10 kW estimate SOFC control frameworks. The oxidant gas is air or oxygen at the SOFC cathode, and the electrochemical decrease of O_2 requires a progression of basic responses and includes the exchange of different electrons. The SOFC cathode must meet the prerequisites of high synergist action for oxygen particle separation and oxygen decrease, high electronic conductivity, substance and dimensional security in conditions experienced amid cell creation and cell operation, warm development coordinate with other cell segments, and similarity and least reactivity with the electrolyte and the interconnection. At long last, the cathode must have a steady, permeable microstructure with the goal that vaporous oxygen can promptly diffuse through the cathode to the cathode/electrolyte interface. These stringent electrochemical and mechanical necessities extraordinarily confine the quantity of appropriate competitor materials. Directing perovskites are the favored cathode materials. Lanthanum manganite ($LaMnO_3$), which, when substituted with low valence components, for example, Ca or Sr, has great p-sort electronic conduction because of the arrangement of huge measure of Mn^{4+}. Additionally, doped $LaMnO_3$ has sufficient electro catalytic movement, a sensible warm development match to YSZ, and soundness in the SOFC cathode working condition. For SOFCs working at generously bring down temperatures, for example, 650–800°C, elective cathode materials, commonly containing progress metals, for example, Co, Fe, and additionally Ni on the B site, have been produced and enhanced for better execution. All in all, these materials offer higher oxide particle dissemination rates and display speedier oxygen lessening energy at the cathode/electrolyte interface contrasted and lanthanum manganite. Notwithstanding, the warm extension coefficient of cobaltite's is significantly higher than that of the YSZ electrolyte, and the electrical conductivities of ferrites and nick elites are low. By and by, promising outcomes have been accounted for utilizing these materials, however as a rule the enhanced cathodic execution is found to diminish amid the cell lifetime because of substance or microstructural unsteadiness. Minimization of cathodic polarization misfortunes is one of the greatest difficulties to be overcome in getting high, stable influence densities from bring down temperature SOFCs. Perovskite sort oxides, especially (La, Sr) MO_3 (M = Co, Fe, or potentially Ni), are exceptionally receptive toward YSZ. Along these lines, a thin layer, for the most part of a CeO_2 based material, is utilized to decrease the concoction response between the cathode and YSZ. Microstructure additionally assumes a noteworthy part in the cathode polarization; this is especially evident when a composite cathode e.g., Sr doped $LaMnO_3$ + YSZ, which demonstrates a superior execution contrasted with a solitary structure cathode, is utilized. Leather expert et al. have demonstrated that polarization protection relies on the grain size of the ionic conductor in the composite terminal and the volume division of porosity. The anode must be a fantastic impetus for the oxidation of fuel (H_2, CO), stable in the lessening condition of the fuel, electronically leading, and should have adequate porosity to enable the vehicle of the fuel to and the vehicle of the results of fuel oxidation far from the electrolyte/anode interface where the fuel oxidation response takes place. 7 alternate necessities incorporate coordinating of its warm development coefficient with that of the electrolyte and interconnect; trustworthiness of porosity for gas saturation; concoction strength with the electrolyte and interconnect; and materialness to use with adaptable powers and debasements. What's more, taken a toll viability is dependably a factor for commercialization. Ni-YSZ cermet's are the most usually utilized anode materials for SOFCs. Ni is an astounding impetus for fuel oxidation; be that as it may, it has a high warm development coefficient (13.4×10^{-6}/°C), and displays coarsening of microstructure because of metal accumulation through grain development at cell operation temperatures. YSZ in the anode compels Ni conglomeration and averts sintering of the nickel particles, diminishes the viable warm development coefficient conveying it nearer to that of the electrolyte, and gives better attachment of the anode with the electrolyte. In the Ni/YSZ cermet anode, nickel has double parts of the impetus for hydrogen oxidation and the electrical current conduit. What's more, it is additionally profoundly dynamic for the steam changing of methane. This reactant property is misused in the supposed inside transforming SOFCs that can work on energizes made out of blends of methane and water. In spite of the fact that nickel is a magnificent hydrogen oxidation and methane steam reforming impetus, it additionally catalyzes the development of carbon from hydrocarbons under diminishing conditions. Unless adequate measures of steam are available alongside the hydrocarbon to expel carbon from the nickel surface, the anode might be pulverized. Thus, notwithstanding when utilizing methane as the fuel, generally high steam-to-carbon proportions are expected to stifle this harmful response. Sadly, because of the high synergist movement of nickel for hydrocarbon splitting, this approach does not work for higher hydrocarbons, and it is by and large unrealistic to work nickel-construct anodes in light of higher hydrocarbon-containing powers without pre-improving with steam or oxygen. Regardless of this disadvantage, Ni-YSZ cermet remains the most usually used anode material for SOFCs and is acceptable for cells working on perfect, changed fuel. Be that as it may, progressed SOFC plans put extra limitations on the anode, for example, resilience of oxidizing conditions as well as the capacity to endure huge amounts of sulfur or potentially hydrocarbon species in the fuel stream. Elective materials, for example, ceria or strontium titanate ceria blends, have yielded some encouraging outcomes in these outlines, yet the advantages got as far as sulfur, hydrocarbon and additionally redox resilience are offset different confinements, for example, the trouble of incorporating such materials with existing cell and stack creation procedures and materials.

Copper cermet anodes have additionally been proposed for intermediate temperature which is less than 800°C. SOFCs expected to work specifically on hydrocarbon fills without earlier reformation, yet the absence of reactant action for oxidation of fuel in copper and sintering of copper at the phone working temperatures have restricted their utilization in pragmatic SOFCs. Fired anodes, for example, doped perovskites, (La, Sr) TiO_3, (La, Sr) CrO_3, with expansion of CeO_2 have likewise been explored as anodes, yet they experience the ill effects of higher electronic protection.

12.1.2. Application

Utilizing planar SOFCs, stationary power age frameworks of from 1 kW to 25 kW estimate have been manufactured and tried by a few associations. A few hundred 1 kW size joined warmth and power (CHP) units for private application were field tried by Sulzer Hexis; notwithstanding, their cost and execution corruption was high and stack lifetime too short. With enhanced fixing materials and fixing ideas, planar SOFC model frameworks in the 1 to 5 kW sizes have as of late been created and are being tried by different associations with more noteworthy achievement. Utilizing tubular (tube shaped) SOFCs, Siemens, and Westinghouse fabricated a 100 kW climatic power age system. The framework was effectively worked for a long time in the Netherlands on desulfurized gaseous petrol with no perceivable execution corruption. It gave up to 108 kW of air conditioning power at a productivity of 46% to the Dutch lattice and roughly 85 kW of high temp water for the nearby region warming framework. At the finish of the operation in The Netherlands, the framework was moved to a German utility site in Essen, Germany, where it worked effectively for an additional 4,000 hours. Subsequent to supplanting a few cells, the framework was then introduced and worked in Italy, for more than two years, again with extremely stable execution. Siemens and Westinghouse tubular cells have likewise been utilized to create and field test over twelve 5 kW measure CHP units, each about the span of a cooler. These units gave magnificent execution and execution soundness on an assortment of hydrocarbon powers. Nonetheless, at exhibit, their cost is high; future such units are relied upon to utilize higher power thickness substitute tubular geometry cells to drive down the cost.

Another utilization of SOFC frameworks is in the transportation division. The polymer electrolyte film (PEM) energy unit is for the most part viewed as the power module of decision for transportation applications. PEM power devices require unadulterated H_2, with no CO, as the fuel to work effectively. Notwithstanding, by and by no H_2 framework exists, and on-board reformer frameworks to create H_2 from existing fuel base (gas, diesel) are actually testing, complex, and costly. Moreover, it is hard to dispense with the CO totally from the reformate stream. Interestingly, SOFCs can utilize CO alongside H_2 as fuel, and their higher working temperature and accessibility of water on the anode side makes on-cell or in-stack reconstruction of hydrocarbon energizes achievable. Likewise, no respectable metal impetuses are utilized as a part of SOFCs diminishing expense of the cells. The underlying use of SOFCs in the transportation part will be for on-board helper control units (APUs). Such APUs, working on existing fuel base, will supply the ever-increasing electrical power requests of extravagance cars, recreational vehicles, and overwhelming obligation trucks. Delphi Corporation has built up a 5 kW APU utilizing anode-upheld planar SOFCs. This unit is planned to work on fuel or diesel, which is changed through synergist halfway oxidation. The building pieces of such an APU comprise of a SOFC stack, fuel reconstruction framework, squander vitality recuperation framework, warm administration framework, process air supply framework, control framework, and power hardware and vitality stockpiling (battery) framework. Delphi has diminished the mass and volume in progressive age APU units to meet the stringent car prerequisites; the rest of the issues of startup time and resilience to warm cycling are by and by being chipped away at.

12.2. PHOSPHORIC ACID FUEL CELL (PAFC)

Energy units, which utilize phosphoric corrosive arrangement as the electrolyte, are called phosphoric corrosive power devices (PAFCs). As Eq. (12.1) shows the phosphoric corrosive in fluid arrangement separates into phosphate particles and hydrogen particles, the hydrogen particles (H^+) go about as the charge transporter.

$$H_3PO_4 \rightarrow H \quad + H_2PO_4 \qquad (1)$$

Phosphoric corrosive is artificially steady, and is anything but difficult to deal with. It additionally has an amazingly low vapor weight even at a working temperature of 200°C (473 K). This suggests phosphoric corrosive in the electrolyte layer can't be effectively released from the energy component together with the cell fumes gas, albeit even such moment release, brings about the debasement of cell execution in the long haul.

A reasonable working standard is portrayed through equations. At the fuel anode, unadulterated hydrogen or reformate fuel gasses the central part being hydrogen is provided, and air is provided at the air terminal; the subsequent electrochemical response yields an electric power yield. At the fuel terminal, hydrogen responds at the anode surface to end up hydrogen particles and electrons, and the hydrogen particles move toward the air cathode inside the electrolyte.

$$Fuel\,electrode: H_2 \rightarrow 2H^+ + 2e^- \quad (2)$$

At the air terminal, the hydrogen particles, which have relocated from the fuel cathode; electrons, which have gone through the outer circuit, and oxygen provided from outside, consolidate to deliver water in the accompanying response

$$Air\ electrode: \frac{1}{2}O_2 + 2H^+ + 2e^- \rightarrow H_2O \quad (3)$$

Hence the net fuel cell reaction produces water as follows:

$$H_2 + \frac{1}{2}O_2 \rightarrow H_2O \quad (4)$$

12.2.1. Structure and Features

The PAFC itself comprises of a couple of permeable terminals (the fuel cathode and air anode) framed from predominantly carbon material, between which is set an electrolyte layer comprising of a lattice impregnated with exceptionally thought phosphoric corrosive arrangement. The synergist layer of the terminals where responses occur comprises of the carbon material, minute metal impetus particles, and water repellant material, in a development with the end goal that the response gas is provided and the electrolyte held viably [91].

The voltage acquired from a solitary energy unit is from 0.6 to 0.8 V or thereabouts, in genuine power plant a few hundred cells are stacked and associated in arrangement, framing a sub unit called a "phone stack." Heat is produced because of vitality misfortunes over the span of the electrochemical response of hydrogen with oxygen, thus cooling plates are embedded at customary interims between power modules, and cooling water is gone through them to keep up a phone working temperature of around 200°C (473 K).

The PAFC don't endure the carbon dioxide-prompted electrolyte degeneration seen in soluble energy units, thus can utilize transformed gas got from non-renewable energy sources, however costly platinum impetus is essential keeping in mind the end goal to advance the terminal responses.

In this way, it can influence utilization of city to gas (flammable gas based) and other existing fuel foundation. Nonetheless, when CO exists at high fixations, as in coal-gasified gas, the platinum impetus utilized as a part of anodes is harmed, prompting execution debasement, so utilization of such powers is illogical without viable methods for wiping out CO. This gives an extra requirement.

The working temperature is around 200°C (473 K). Subsequently if the cell is planned to such an extent that it doesn't reach the phosphoric corrosive, copper, press, and different metals can be utilized. Likewise, so as to invest the anode impetus layer with water-repulsing properties, a fluoride gum (PTFE) or other exceptionally warm safe natural material may likewise be utilized. With a specific end goal to expel the warmth produced by the anode responses, the energy component is itself water-cooled as said above.

Squander warm at a temperature run underneath 200°C is accessible; which can't simply be utilized for space warming and water warming, however can likewise be removed to some degree as steam and utilized as the warmth wellspring of refrigeration hardware for cooling. The electric power age proficiency of PAFCs under air weight operation is around 40% (LHV premise), which is prevalent, or possibly aggressive with existing gas turbine and gas motors. Properties of low Nox and low commotion make them appropriate for cogeneration frameworks for urban naturally well-disposed power sources.

Not at all like the high temperature energy unit frameworks, for example, MCFCs, and SOFCs, a joined cycle framework with gas turbine or steam turbine generators to expand the framework effectiveness is for the most part troublesome for PAFCs, since the nature of plant fumes warm is lacking for such purposes.

In pressurized PAFC frameworks, thought reformer fumes gas at raised weight and temperature can be gone through an expanded to drive an air compressor or an electric power generator, the aggregate power age proficiency remains in a scope of 44–46% that is LHV premise.

12.2.2. Operation and Working

The working temperatures and corrosive convergences of PAFCs have expanded to accomplish higher cell execution; temperatures of around 200°C and corrosive groupings of 100% H_3PO_4 are regularly utilized today. In spite of the fact that the present practice is to work at climatic weight, the working weight of PAFCs outperformed 8 atm in the 11 MW electric utility show plant, affirming an expansion in control plant effectiveness. Be that as it may, various issues remain whether to outline and work PAFC units at environmental versus pressurized conditions. Basically, little, multi-kW PAFC control units that were the concentration of introductory business applications prompted air weight operation. In spite of the fact that pressurization expanded effectiveness bring down fuel cost, it convoluted the power unit – bringing about higher capital cost. The financial exchange off favored less complex, barometrical operation for early business units. Another essential issue, free of energy unit measure, is that weight advances consumption. Phosphoric corrosive electrolyte H_3PO_4 produces a vapor. This vapor, which shapes over the electrolyte, is destructive to cell areas other than the dynamic cell territory. These cell areas are at a blended voltage open circuit and cell voltage that can be over 0.8V/cell. That is the utmost above which consumption happens dynamic zone constrained to operation under 0.8 V/cell. An expansion in cell add up to weight causes the halfway weight of the H_3PO_4 vapor to expand, causing expanded erosion in the cell. Cell temperature should likewise be expanded with pressurized conditions to deliver steam for the steam reformer [92].

Carbon dark and graphite were adequately steady to supplant the more costly gold-plated tantalum cell equipment utilized at the time. The utilization of high-surface zone graphite to help Pt allowed an emotional decrease in Pt stacking without relinquishing terminal execution. It was accounted for that "without graphite, a sensibly modest corrosive energy component would be inconceivable, since no other material joins the fundamental properties of electronic conductivity, great consumption protection, low thickness, surface properties (particularly in high territory frame) and, most importantly, minimal effort." In any case, carbon erosion and Pt disintegration turn into an issue at cell voltages above ~0.8 V. Thus, low current densities at cell voltage over 0.8 V and hot sit without moving at open circuit potential ought to be maintained a strategic distance from. The permeable anodes contain a blend of electro-impetus upheld on carbon dark and a polymeric folio, generally PTFE (30 to 50 wt %). The PTFE ties the carbon dark particles together to shape a vital, however permeable, structure that is upheld on a permeable graphite substrate. The graphite structure fills in as a help for the electro-impetus layer, and additionally the present authority. A run of the mill graphite structure utilized as a part of PAFCs has an underlying porosity of around 90%, which is decreased to around 60 % by impregnation with 40 wt % PTFE. This wet-verification graphite structure contains macro pores of 3 to 50 μm distance across, middle pore breadth of around 12.5 μm and micro pores with a middle pore width of around 34 Å for gas penetrability. The composite structure, comprising of a carbon dark/PTFE layer on the graphite substrate, frames a steady, three-stage interface in the energy component, with H_3PO_4 electrolyte on one side (electro-impetus side) and the reactant gas condition on the other. A bipolar plate isolates the individual cells and electrically associates them in arrangement in an energy component stack. In a few outlines, the bipolar plate additionally contains gas channels that sustain the reactant gasses to the permeable anodes and evacuate the response items and inert. Bipolar plates produced using graphite gum blends that are carbonized at low temperature of about 900°C are not appropriate due to their quick debasement in PAFC working conditions. Be that as it may, consumption dependability is enhanced by warm treatment to 2,700°C, i.e., the erosion current is diminished by two requests of extent at 0.8 V in 97 % H_3PO_4 at 190°C and 4.8 atm. The all-graphite bipolar plates are adequately erosion safe for an anticipated existence of 40,000 hours in PAFCs, however they are still generally expensive to create. An average PAFC stack contains cells associated in arrangement to get the down to earth voltage level wanted for the heap. In such a course of action, singular cells are stacked with bipolar plates between the cells. The bipolar plates utilized as a part of early PAFCs comprised of a solitary bit of graphite with gas channels machined on either side to coordinate the stream of fuel and oxidant. At present, both bipolar plates of the past plan and new outlines comprising of a few segments are being considered. In the multi-part bipolar plates, a thin impenetrable plate isolates the reactant gasses in neighboring cells in the stack, and separate permeable plates with ribbed channels are utilized to coordinate gas stream. In a cell stack, the impenetrable plate is subdivided into two sections, and every go along with one of the permeable plates. The electrolyte vaporizes with the goal that a bit of H_3PO_4 escapes from the cell noticeable all-around stream after some time. An electrolyte store plate (ERP), made of permeable graphite, gives enough electrolyte to accomplish a 40,000-hour cell all-consuming purpose (there is no electrolyte substitution) [93]. The ERP likewise suits increments in electrolyte volume because of an expansion in H_2O, so the permeable graphite anodes don't surge. These variances in electrolyte volume happen amid start-up and amid transient operation. The permeable structure, which permits quick gas transport, is likewise used to store extra corrosive to recharge the supply lost by vanishing amid the cell working life. In PAFC stacks, arrangements must be incorporated to evacuate warm produced amid cell operation. Practically speaking, warm has been expelled by either fluid (two-stage water or a dielectric liquid) or gas (air) coolants that are directed through cooling channels found (for the most part about each fifth cell) in the phone stack. Fluid cooling requires complex manifolds and associations, yet preferred warmth expulsion is accomplished over with air-cooling. The upside of gas cooling is its straightforwardness, unwavering quality, and generally minimal effort. Be that as it may, the measure of the cell is restricted, and the air-cooling sections must be significantly bigger than the fluid cooling entries.

12.2.3. Future of PAFC

Phosphoric corrosive anode/electrolyte innovation has achieved a level of development at which engineers submit assets for business limit, multi-unit exhibits and pre-prototype establishments. UTC Fuel Cells has 25 (200 kW) barometrical weight control plants that have worked between 30,000 to 40,000 hours. Most cell parts are graphite, and there has been no electrolyte substitution over the cell life of 40,000 hours. Matrix free units experience broad cycling. Cell parts are fabricated at scale and in huge amounts, showing certainty that anticipated execution will be met. Be that as it may, additionally increments in control thickness and lessened cost are expected to accomplish monetary intensity with other vitality advances, as communicated in the mid-1990s. Energy component engineers keep on addressing these issues. In 1992, UTC Fuel Cells' forerunner, International Fuel Cells, finished a legislature supported, propelled water-cooled PAFC advancement undertaking to enhance the execution and lessen the cost of the two of its air and pressurized innovation for both on location and utility applications. The venture concentrated on five noteworthy exercises [94]

1. create a calculated plan of an extensive stack with an objective of 0.188 W/cm^2, 40,000 hour helpful life, and a stack cost of under \$400/kW.
2. test pressurized Configuration "B" single cells created in a past program, yet enhanced with exclusive outline progresses in substrates, electrolyte repository plates, impetuses, seals, and electrolyte network to show the 0.188 W/cm^2 control thickness objective 2-test a pressurized short stack with subscale measure, enhanced segment cells, and extra upgrades in the fundamental separators and coolers to affirm the stack plan.

3 test a pressurized short pile of enhanced full-measure cell segments, ostensible 1 m^2 estimate, to show the 0.188 W/cm^2 control thickness objective.

4 test a progressed environmental "on location" control unit stack with the enhanced segments.

12.3. POLYMER ELECTROLYTE MEMBRANE FUEL CELL (PEMFC)

Another way to deal with explain the operation and control of Polymer Electrolyte Membrane (PEM) energy components is being produced. A worldwide reactor building approach is connected to PEM power modules to distinguish the basic material science that oversee the progression in PEM energy units. Response building standards are utilized to build up a one-dimensional differential PEM energy unit reasonable for explaining the dynamic execution of PEM cells under all around characterized conditions. PEM power modules utilize a polymer film with corrosive side gatherings to lead protons from the anode to cathode. Water administration in the energy component is basic for PEM power device operation. Adequate water must be ingested into the layer to ionize the corrosive gatherings; overabundance water can surge the cathode of the energy component lessening energy component execution constraining the power yield. A schematic of a polymer electrolyte film hydrogen-oxygen power module is appeared in Figure 12.1.

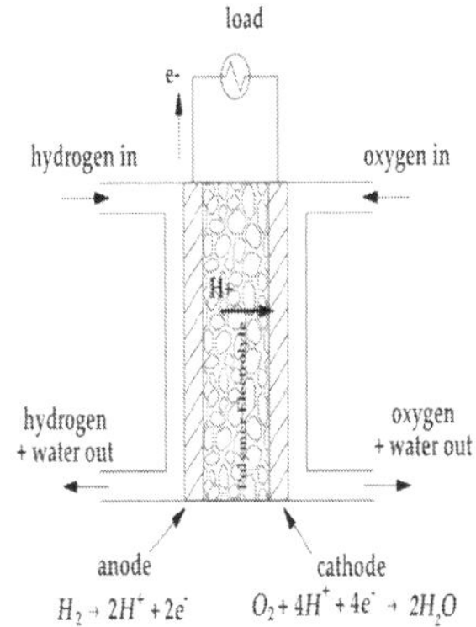

Figure 12.1: Hydrogen oxygen fuel cell. (Source: *http://www.princeton.edu/~benziger/PEMFC_files/image001.gif*).

Hydrogen-oxygen PEM energy component. Hydrogen particles dissociatively adsorb at the anode and are oxidized to protons. Electrons go through an outside load protection. Protons diffuse through the PEM under an electrochemical angle to the cathode. Oxygen particles adsorb at the cathode, are decreased and respond with the protons to create water. The item water is retained into the PEM, or dissipates into the gas streams at the anode and cathode [95].

Proton conductivity in Na ion, and most other polymer electrolytes, increments with water action, and is augmented when equilibrated with fluid (water action aw=1). Operation at aw=1 limits the layer protection for proton conduction, however brings about water buildup which hinders mass exchange to the anodes. Our exploration is centered around finding ideal operational and control plans for water administration in PEM power modules.

The PEMFC (Polymer Electrolyte Membrane Fuel Cell/Proton Exchange Membrane Fuel Cell) are power devices where the electrolyte is made of a natural polymer that has the normal for a decent proton bearer when in nearness of a water arrangement.

The main fluid existent in this sort of cell is water and, subsequently, the erosion issue is least. The water administration in PEMFC cells is critical. The way that the film should be dependably in water arrangement, will constrain the temperature of the cell operation to the water vaporization temperature, keeping away from layer dryness. Not permitting vitality co-age from the warm vitality dispersed. The proficiency of the entire procedure is around half to 60%.

The principle starting trouble to the utilization of PEMFC cells was the high cost of the platinum fundamental as an impetus in the anodes. At first the PEMFC cells utilized 4 mg of platinum for every cm^2 of film region that made costly the utilization of this sort of cell. These days, with the mechanical improvement in the materials for the cathodes it is conceivable the utilization of 0.15 $mg\backslash cm^2$ of platinum, with the same electrical productivity, making the power device more affordable and nearer to the sparing viability [96].

The PEMFC's have a vigorous outline and are generally easy to manufacture. The primary attributes are the low working temperature that is about 80°C that permits a quick reaction while being turned on and off. To these properties can be included the benefits of the immediate use of air in the cathode and zero emanation of CO_x and NO_x, making this sort of cell the most encouraging option for the substitution of the burning motors utilized as a part of transportation vehicles, and for the electrical vitality creation in little and medium size stationary units underneath 250 kW [97].

The quantity of energy units in the cluster decides the estimation of the operation voltage in a power device stack. The voltage is additionally characterized by the inside misfortunes, materials, amassing strategies, the present power forced by the heap and the entire operational process. In this way, the electric power created by the stack is characterized by a progression of collecting and operational parameters.

12.4. ALKALINE MEMBRANE FUEL CELL (AFC)

The first passage in the AMFC zone was distributed in 2005, since when movement and intrigue have kept on expanding steeply globally. Zeng and Varcoe have as of late investigated the creating patent writing. A few scientists have as of late named these frameworks HEMFCs, hydroxide trade film power devices; that phrasing isn't completely fitting in perspective of the mind-boggling hydroxide/hydrogen carbonate/carbonate equilibria that are available (even in the wake of dealing with layers in air), unavoidable in the utilization of air (containing CO_2(g)) as wellspring of oxidant and furthermore delivered in the oxidation of methanol in coordinate methanol energy components (DMFCs). Another acronym connected to these frameworks, APEMFCs, is additionally conceivably confounding, the first three expressions of the full shape at that point apparently being "soluble proton trade" following the fundamentally the same as PEMFC acronym, yet being decipherable on the other hand as antacid polymer electrolyte layer power modules, perhaps AEMFC would be clearer. Soluble energy components AFCs, hydrogen fueled cells with a basic fluid electrolyte, for example, KOH(aq) are the best performing of all known customary hydrogen-oxygen power devices operable at temperatures beneath 200°C. This is because of the simple energy at the cathode and at the anode; less expensive non-respectable metal impetuses can be utilized, for example, nickel and silver, decreasing expense. Mc Lean et al. gave complete survey of basic power module technology [98].

The real reason for the debasing execution of AFCs is the ensuing precipitation of metal carbonate precious stones (most usually Na_2CO_3 or K_2CO_3, contingent upon the antacid electrolyte utilized) in the electrolyte-filled pores of the terminals, blocking pores and mechanically disturbing and wrecking dynamic layers. The cost of energy units still retards commercialization in many markets. AFCs are promising on a cost premise for the most part in light of the fact that shabby and generally plentiful non-platinum-assemble metals (non-PGM) are feasible impetuses, yet are frustrated by corruption because of development of hastens as above. Impetus electro energy for fuel oxidation and oxygen diminishment are enhanced in soluble, instead of acidic, conditions (the corrosive steadiness rule blocks the utilization of most non-PGM impetuses in PEMFCs). The substitution of the KOH(aq) electrolyte with a basic electrolyte layer (AEM), to give AMFCs, holds the electro reactant focal points however presents CO_2 resistance as there being no portable cations that could give carbonate/bicarbonate encourages with the extra preferred standpoint of being an all-strong state power module as with PEMFCs that is, no leaking out of KOH(aq). Also, thin (low electronic protection and effectively stamped and shabby metal mono/bipolar plates can be utilized, with diminished consumption determined issues at high pH and the cost of bi polar plates for PEMFCs can be as much as 33% of the cost of the stacks themselves. A key but then to be convincingly met necessity is, be that as it may, the improvement of a dispersible soluble ionomer here and there named an ionomer to expand ionic contact between the impetus response destinations and the particle conductive films. As on account of AFCs, water is created at anode and devoured at cathode in AMFCs (when fueled with hydrogen and with four electrons lessening of oxygen at the cathode), which is on a very basic level diverse to what happens in PEMFCs containing acidic electrolytes; this can cause high overpotentials at AMFC anodes, because of suspected flooding. The utilization of an AEM as a strong electrolyte including no metal cations averts precipitation of carbonate or bicarbonate salts. The electrolyte containing the cationic gatherings is as of now a strong. The carbonation procedure is fast regardless of whether the AEM has been presented to the air for just a brief span. The conductivities of the AEMs in OH shape may have been thought little of on the grounds that most examinations to date have not unveiled energetic CO_2 prohibition strategies amid conductivity estimations. It has been conjectured that OH particle conductivities in AEMs can be assessed by measuring the ionic conductivities of HCO_3 from AEMs and duplicating by 3.8. This carbonation procedure may not to be a difficult issue due to an in situ "self-purging system" since OH anions are persistently created at the cathode in AMFCs. AEMs are strong polymer electrolyte films that contain positive ionic gatherings commonly quaternary ammonium (QA) practical gatherings, for example, poly NMe^{3+} and portable contrarily charged anions. A broadly cited worry with AEMs is layer steadiness, particularly at raised temperatures. The general issues are [99]:

1. The dispersion coefficients and mobilities of OH anions are not as much as that of H+ in most media, and QA ionic gatherings are less separated than the run of the mill sulfonic corrosive gatherings (pKa for sulfonic corrosive gatherings are typically1 however for QA bunches the related pKb esteems are around +4); there were worries that AEMs would not have sufficient characteristic ionic conductivities for application in power modules.
2. The OH anions are viable nucleophiles which possibly cause corruption through a direct nucleophilic removal or potentially (b) a Hofmann end response when a b-hydrogen is available; methyl (CH_3) gatherings may likewise be dislodged by OH particles shaping tertiary amines and methanol. In the event that the AEMs contain great leaving gatherings (e.g., QA – NMe^{3+} gatherings) at that point the substance solidness of the AEMs may have been deficient for use in power devices, especially at raised temperatures.
3. Forerunner anion-trade layers are by and large submerged in watery NaOH/KOH answers for trade them to the OH frame AEM; the AEM must have the synthetic solidness to withstand this procedure. In spite of this, over 10 years back, the strong qualities of different benzyl-trimethyl-ammonium-based AEMs were observed to be steady at up 75C in NaOH(aq) at focuses up to 6 mol/dm^3 for a few days.

A noteworthy potential use of AMFCs is, in any case, as power hotspots for at or close room temperature (with respect to PEMFCs), which means such debasement can be negligible. This passage considers the present comprehension and use of AEMs in hydrogen-fueled AMFCs and other power device sorts utilizing AEMs. it empowers examination between hydrogen or methanol fueled PEMFCs and AMFCs; the anode responses in a hydrogen fueled AMFC are talked about above. Rather than PEMFCs, operation of an AMFC requires the nearness of water as a reagent at the cathode (oxygen diminishment response, ORR – to frame OH) and the item water is shaped by the hydrogen oxidation response (HOR) at the anode (instead of being framed at the cathode in PEMFCs). The section starts with thought of the principle classes of AEMs.

12.4.1. Properties

The most well-known class of NaOH or KOH free AEMs being examined for use in energy components depends on QA science and has sensible strength in basic situations (particularly those AEMs containing benzyl trimethyl ammonium trade locales). There are three fundamental classes of substance corruption response systems by which nucleophilic OH anions can expel QA gatherings. The nearness of different hydrogens permits the Hofmann disposal response to happen, frequently in parallel to the responses talked about underneath, yielding alkene (vinyl) gatherings; this can offer ascent to QA, AEMs that have low warm and substance (to salt) secure qualities. On the off chance that, as in RGAEMs, no b-hydrogen iotas are available, coordinate nucleophilic substitution responses were generally thought to happen yielding liquor and tertiary amine gatherings. In any case, late thickness utilitarian hypothesis (DFT) counts and deuterium trade tests at Los Alamos National Laboratory (USA) on demonstrate little QA containing mixes show that a system including a ylide middle of the road (trimethyl ammonium methylide, otherwise called a 1,2-dipolar ylide compound) may prevail and be more serious when the AEMs are dried out. AEMs from Tokuyama Co have great thermochemical steadiness. The thin (10 mm) Tokuyama "energy unit review" AEMs (A010, A201 – once A006 and A901) and their formative dispersible soluble ionomers (A3 ver.2 and AS– 4) have additionally been tried at up to 50 C by a few research bunches in coordinate liquor power devices [100]. Quaternized pyridinium-or phosphonium-based AEMs were thought to have thermochemical strong qualities that are not reasonable for use in AEM power modules but rather there are reports of late work on polysulfone-phosphonium-based AEMs and anionomers (see later). Surrey's benzyl trimethyl ammonium-containing S80, S50, and S20 RG–AEMs (the number assigning the completely hydrated thicknesses in micrometers) are synthetically steady up to 80°C in watery KOH (aq, 1 mol/dm^3) and can display ionic conductivities > 0.03 Scm^{-1} at room temperature when completely hydrated; the excitant in situ ionic conductivity of completely hydrated S80 is 0.06 S cm^{-1} at 60°C (cf. Nafion acidic PEMs are ordinarily is greater than 0.1 s/cm^{-1} at these temperatures). AEM conductivities are, in any case, extensively diminished at humidities RH/% < 100 and drop to values between 0.01 and 0.02 s/cm^{-1} after just a hour when presented to air (particularly with thin layers) because of the response of OH anions with CO_2 framing CO_2 and HCO_3 inside the film. The lower separation steady for NMe_3OH gatherings requiring a higher number of water atoms for finish separation), contrasted with SO_3H bunches in PEMs, and a low number of water particles specifically connected with the ionic gatherings prompt the poor execution at low RHs at high humidities a great part of the water show in AEMs is situated in totals not straightforwardly connected with the ionic gatherings. The creating distributed and patent writing here has as of late been surveyed by Zeng and Varcoe. Additionally classes of AEM for energy units have included films in view of quaternized poly (epichlorohydrin), polysulfone, poly (phthalalazinone ether ketone), poly (2,6-dimethyl-1,4–phenylene oxide), and poly(vinyl liquor) joined with (2,3-epoxypropyl) trimethyl ammonium chloride. Regular quaternizing operators incorporate alkyl iodides, trialkyl amines, N,N,N,'N'– tetramethyl alkyl-1, n-diamines, polyethyleneimine, 1,4-diazabicyclo-[2.2.2]-octane (DABCO), and 1-azabicyclo-[2.2.2]-octane. The final two of these have been utilized widely; they contain b-hydrogen yet their structures don't allow sub-atomic compliances supported in the Hofmann disposal component. Metal-cation-containing AEMs in view of doping polymer films (e.g., polybenzimidazole (PBI), poly (vinyl liquor) and its composites, and biocompatible chitosan) with NaOH or KOH(aq) are additionally being researched, however the nearness of versatile cations may present issues related with precipitation of carbonate salts. Ionic conductivities of AEMs are by and large lower than those of practically identical PEMs. This isn't shocking as the arrangement versatility of OH is 33% to one portion of that of a H^+ (contingent upon nature and if there mo104/cm^2 V_1 s_1 = 20.64 for OH (aq) anions and 36.23 for H^+(aq) cations at 298 K. One technique for upgraded ionic conductivities is to build the particle trade limit (IEC) by means of engineered strategy, yet this regularly prompts an abatement in the mechanical quality because of unnecessary water takes-up. Another procedure is to combine custom fitted films that will display hydrophilic(ionic)– hydrophobic (nonionic) stage isolation and constant ionic areas, which is estimated to build ionic conductivities. Late concentrated investigations have, be that as it may, been accounted for to prompt AEMs with high ionic conductivities, apparently tantamount to Nafion. These promising AEMs are still to be assessed in AMFCs. Most hydrocarbon AEMs are dissolvable in different solvents, which is conceivably valuable for the detailing of soluble ionomers required for the readiness of superior layer anode congregations (MEAs). On the off chance that the conductive properties announced can be converted into high power yields, at that point AMFC exhibitions similar to those of PEMFCs can be normal sooner rather than later [101].

12.4.2. Applications

The utility of AEMs as potential electrolytes in power devices emerges not just from the prospects for the utilization of non Pt aggregate metal (non-PGM) impetuses and less expensive energy unit segments less destructive condition, yet in addition from potential for utilization of option fills. Alcohols and diols, sodium borohydride ($NaBH_4$), and hydrazine (H_2NNH_2) have all been utilized specifically as powers in AMFCs. Licenses presented more energizes including hydrated hydrazine ($NH_2NH_2{\cdot}H_2O$), hydrazine carbonate ($(NH_2NH_2)_2CO_2$), hydrazine sulfate ($NH_2NH_2{\cdot}H_2SO_4$), monomethyl hydrazine (CH_3NHNH_2), smelling salts (NH_3), hetero cycles such as imidazole and 1,3,5-truazineand3-amino-1,2,4-triazole, and hydroxy lamines, for example, hydroxylamine (NH_2OH) and hydroxylamine sulfate ($NH_2OH{\cdot}H_2SO_4$); the impetuses were co-based for the fuel side (anode) and Ag/C, Pt/C and Ni/C for the oxygen decrease side (cathode). EP2133946 uncovered the utilization of change metals as impetuses in AMFCs. There are various reports on the utilization of non-Pt catalyst, such as MnO_2, Ag/C, Au/C, FeTPP/BPC (Black Pearl Carbon), CoPPyC, FeCoCNF (Carbon Nanofiber), CoFeN/CHLH [84] in AMFC cathodes and Cr-beautified Ni/C, Fe-Co-Ni/C at anodes. Zhuang et al. directed a possibility examination of the utilization of AEMs in DMFCs; this critical investigation focused on the thermodynamic burdens versus motor favorable circumstances of such cells by considering the responses occurring in watery arrangements and the creators presumed that the thermodynamic voltage misfortunes because of the pH contrast over the AEM would be around offset by the dynamic voltage picks up. Ogumi et al. considered direct liquor power modules (methanol and ethylene glycol) with a business AEM (Tokoyama AHA, thickness 240 mm). Arrangements of ethylene glycol and methanol were made to a grouping of 1 mol/dm^3 broke down in KOH (aq, 1 mol/dm$_3$), thus the framework was not sans metal cation (basic with different AMFCs getting sensible power levels with alcohols as fills). Cell voltages were around 100 mV higher for ethylene glycol contrasted with methanol. This power device isn't exactly a soluble simple of a corrosive layer DMFC, as the creators utilized basic fuel arrangement. No sign was given concerning fulfillment of oxidation of the ethylene glycol to CO_2 and H_2O. Yu and Scott likewise detailed the operation of an immediate methanol antacid power module, with platinized titanium work anodes (1.5 mg cm^2 Pt/Ti) and a business soluble anion-trade film (Morgane ADP layer from Solvay SA). It has been recommended that alkali would make a decent vitality vector or transporter and an aberrant fuel for a hydrogen fuel cell; alkali is naturally favorable (it is as of now utilized as a compost and kills corrosive rain), effectively accessible efficiently, contains half more H per dm^3 than fluid hydrogen, and is a fluid at much lower weights (8–9 bar); it likewise has the upside of a solid odor, permitting the less demanding area of holes. Splitting alkali to shape hydrogen free of hints of smelling salts for use in conventional PEMFCs requires temperatures of over 900°C; PEMFCs can't endure any NH3 pollution. Leeway of AFCs is that they will endure low levels of NH_3, thus splitting can be embraced at bring down temperatures; this still can't seem to be investigated utilizing AMFC approaches [102].

12.4.3. Future Developments

The planning of pertinent AEMs includes a trade-off between the properties of the film, for example, the concoction and warm soundness, particle trade limit (IEC), particle conductivity, mechanical properties, water take-up, and dimensional dependability. When all is said in done, basic anion-trade polymer electrolytes can be polymerized straightforwardly from functionalized monomers, polymerized from monomers with ensuing functionalization or by functionalizing a financially accessible polymer. The foundation of the polymer is generally chosen for its great concoction and warm dependability and, along these lines, commonly incorporates sweet-smelling rings and additionally a level of fluorination: run of the mill polymer classes incorporate poly sulfones and poly etherketones, polyimides, poly(phenylene), poly (phthalazinone ether sulfone ketone), polyepichloro hydrin homo polymer, polybenzimidazole (PBI), poly(phenylene oxide), radiation-joined copolymers, inorganic natural half and halves, and even per fluoronated layers, for example, Nafion. The dynamic utilitarian gatherings are ordinarily quaternary ammonium sort (NR^{3+}) with an unmistakable inclination for trimethylammonium ($N(CH_3)^{3+}$) gatherings (pKa (H_2O) = 9.8). A reasonable AEM will have a high particle trade limit, high ionic conductivity, and thermochemical security, yet will show a low level of swelling on hydration. There are a few general engineered approaches for the readiness of AEMs. Fluorine-containing polymers for the most part indicate higher warm dependable qualities than hydrocarbon polymers. AEMs in light of a poly (arylene ether sulfone) containing fluorine iotas demonstrate high ionic conductivities 63 ms/cm^1 in CO frame at 70°C); this is a key outcome as CO_2 anions have weaken arrangement mobilities that are under 33% of that of OH anions. (It is uncommon to see CO_2 conductivities over 30 ms/cm^1. Illumination of polymer films (and powders, and so on.) utilizing X-beams, g-beams, or electron shafts as at Surrey, see past segment is a flexible approach to present different utilitarian gatherings on the polymer spines. An extensive variety of artificially and warm stable polymers, for example, ETFE and FEP, can be picked as the base films for the generation of AEMs. Furthermore, there is a wide decision of useful monomers accessible that can be utilized to bring particle trade bunches into the united polymeric chains [103]. A typical methodology in the amalgamation of AEMs is to present halogen alkyl bunches onto the spine or side chains of the polymer by means of chloro alkylation, fluorination, bromination, or chlorination, trailed by amination/quaternization and finally particle trade. Very cancer-causing chloro methyl ethers have generally been utilized as the chloro alkylation specialist yet more secure systems have been presented, for instance, the chloro methylation operator is produced in situ. An option technique is the earlier presentation of tertiary amine bunches into the polymer, trailed by quaternization. The simplest manufactured course is to utilize idle polymers (as above) doped with concentrated KOH(aq) as above: polybenzimidazole (PBI), poly vinyl liquor (PVA), composite polymers, for example, PVA or hydroxyapatite (PVA or HAP), quaternized PVA or alumina (QPVA/Al_2O_3), PVA/titanium oxide (PVA/TiO_2), chitosan and cross-connected chitosan, copolymers of epichlorohydrin and ethylene oxide, and cross-connected PVA/sulfosuccinic corrosive (10 wt. % SSA) have all been doped with KOH and utilized as AEMs. Patent US5569559 depicts the utilization of polar polymers (most favored being polyethylene oxide) doped with soluble metal hydroxides, (for example, KOH), antacid earth metal hydroxides or ammonium hydroxides, for example, tetrabutylammonium hydroxide; PBI doped with KOH demonstrated the most elevated ionic conductivity, practically identical to Nafion (a standard acidic proton-trade layer, PEM). These materials could, be that as it may, prompt carbonate hastens. Anion-trade polymers that contain methacrylate, ester, amide, or other carbonyl (C = O twofold bond) utilitarian gatherings demonstrate low secure qualities in salt as these practical gatherings are very responsive to nucleophiles, for example, OH.

12.5. MOLTEN CARBONATE FUEL CELL (MCFC)

The MCFC is a high temperature power device with a liquid carbonate electrolyte working in a temperature scope of 580–700°C. The working temperature interim is constrained by poor conductivity of the electrolyte at bring down temperatures, and by quickened erosion and electrolyte vaporization at higher temperatures. An important favorable position of high temperature energy components is the utilization of modest non-respectable impetuses for anode material. The anode of the MCFC is made of nickel, normally alloyed with Cr or Al for microstructural solidness, and the cathode of the MCFC is made of in-situ lithiated nickel oxide (Figure 12.2).

The electrolyte comprises of an eutectic blend of $Li_2K_2CO_3$ or $Li_2Na_2CO_3$, which is suspended in a separator network comprising of $LiAlO_2$, and halfway in the permeable gas dissemination anodes. The high ionic conductivity at the working temperature enable the separator and cathodes to be thick (0.5–0.6 mm and 0.6–1.0 mm individually) contrasted with low temperature energy units without causing an unsuitable voltage drop, something which is vital for good mechanical steadiness, to forestall gas traverse, and to anticipate Ni-shortening in the electrolyte. MCFC [104] frameworks can be worked in air or in pressurized conditions. Air MCFC frameworks are worked with low current thickness contrasted with other power module advancements, because of high ohmic misfortunes in the electrolyte and high polarization misfortunes at the cathode. Pressurized frameworks are utilized to build the power thickness, which is financially valuable, in any case, expanded disintegration of cathode at high weight is an issue for pressurized frameworks.

The hypothetical proficiency of MCFC is high, up to 60% electrical effectiveness and 90% warm productivity. By and by, business frameworks have exhibited an electrical proficiency of 45–55%. The great waste warmth of the MCFC can be used for joined warmth and power operation (CHP) [18], or for co-age, for example, steam-(ST) or gas-turbine (GT) consolidated cycles in which general electrical efficiencies of 60–70% might be come to. Oxygen is lessened with CO_2 at the cathode, while hydrogen is oxidized at the anode to create water and CO_2. In the electrolyte, carbonate particles give the methods for ionic transport between the cathode and the anode. The cathode and anode responses happen as takes after.

Cathode reaction:

$$\frac{1}{2}O_2 + CO_2 + 2e^- \rightarrow CO_3^{2-} \qquad (5)$$

Anode reaction:

$$H_2 + CO_3^{2-} \rightarrow H_2O + CO_2 + 2e^- \qquad (6)$$

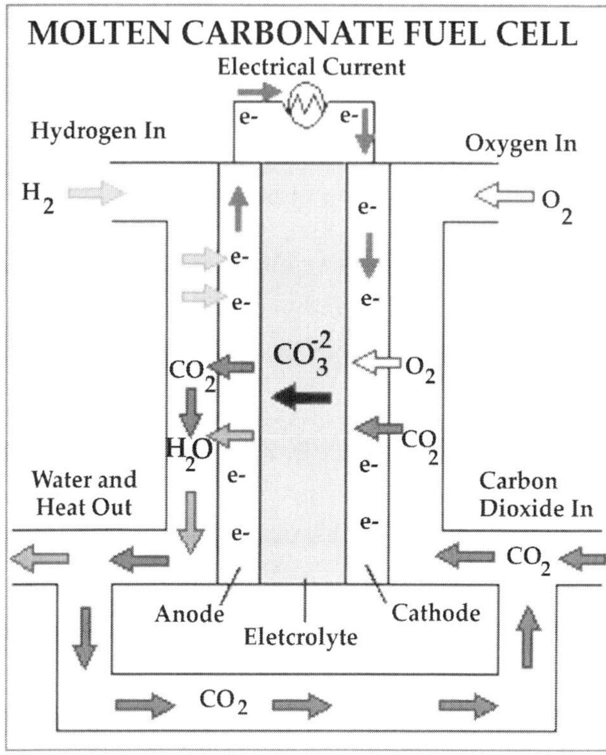

Figure 12.2: Working of MCFC. (Source: *https://i.kinja-img.com/gawker-media/image/upload/s--bJOdo6IW--/c_scale,f_auto,fl_progressive,q_80,w_800/17hk8nbm5r831gif.gif*).

In the subsequent cell response hydrogen and oxygen responds to shape water, and because of a net transport from the cathode to the anode side, CO_2 should be provided to the cathode and additionally oxygen. This is practically speaking settled by recycling a piece of the anode fumes, which is gone through a reactant burner to oxidize remaining fuel in the gas previously it is blended with air at the cathode gulf. The cell response is as per the following [105].

Cell reaction:

$$H_2 + \frac{1}{2}O_2 + CO_2,\ Cat \rightarrow H_2O + CO_2,\ An \tag{7}$$

An extra favorable position of high temperature operation is that CO, which goes about as a toxin in low temperature energy units, might be used as fuel by coordinate CO oxidation at the anode or through the water-gas move response (WGS). The moderate energy of CO oxidation contrasted with hydrogen oxidation, in any case, makes the WGS response more prone to happen.

CO oxidation:

$$CO + CO_3^{2-} \rightarrow 2CO_2 + 2e^- \quad (8)$$

WGS:

$$CO + H_2O \leftrightharpoons H_2 + CO_2 \quad (9)$$

The working temperature of 650°C is adequately high for steam transforming of hydrocarbon powers, i.e., methane, to hydrogen, which makes the MCFC profoundly adaptable regarding fuel supply. Steam transforming is done over an upheld Ni-impetus, and might be performed by outside or inner improving. If there should be an occurrence of the previous, the fuel is changed over to hydrogen and CO_2 before entering the energy component. If there should arise an occurrence of the last mentioned, the steam transforming and the related WGS response happens specifically at the anode.

Steam reforming:

$$CH_4 + H_2O \leftrightharpoons 3H_2 + CO \quad (10)$$

Inner improving uses the warmth delivered by the exothermic hydrogen oxidation for the endothermic steam transforming response, which streamlines warm administration of the stack. The cooling burden might be diminished with as much as half by applying interior improving. Besides, when hydrogen is expended in the anode response, this drives the steam transforming response to the correct which yield higher fuel usage and higher general productivity. In outer improving the warmth frame the synergist burner can used for the changing response. It is utilized for pressurized frameworks, as the steam changing is advanced by high temperature and weight. The water-gas move response is reversible and quick and can be thought to be at balance inside the power device.

Carbon testimony, which is unfavorable for the power device, may happen through the alleged Boudouard response. Humidification of the fuel gas diminishes the danger of carbon statement as it lessens the centralization of CO through the WGS response.

Boudouard reaction

$$2CO\Delta\langle C\rangle + CO_2 \tag{11}$$

The WGS response and the immediate oxidation of CO can influence the OCV by changing the gas organization and by causing a blended potential, as the reversible cell potential for a hydrogen cell and a carbon monoxide cell computed from the estimation of Gibbs free vitality, are somewhat extraordinary.

12.5.1. Applications for Carbon Capture and Storage (CCS)

The idea of MCFC coordinated power age and CO_2 partition adds two essential highlights to the CCS foundation. In the first place the catch of CO_2 can be made at a high fixation; Case considers gauges that the CO_2 focus at the anode fumes can be as high as 80%. Furthermore, the expansion of the MCFC control yield limits the general punishment on plant effectiveness for a practically identical decrease in CO_2 outflows. Besides, the MCFC framework is likewise profited by the excess of anode fumes gas distribution to the cathode gulf, since CO_2 is provided in the vent gas.

In any case, the centralizations of both CO_2 and oxygen in ignition pipe gas are essentially lower than what is utilized as a part of standard operation of the MCFC, which influences the power device execution adversely. As far as possible for CO_2 lean oxidant gas relies upon a huge number of interdepending factors, for example, CO_2 fixation and use factor, current load, stream rates and cell plan. At last, for the utilization of MCFCs as a CCS application, the working conditions and execution are tradeoffs between augmenting

a) the power device yield.

b) the CO_2 expulsion [106].

In framework examination investigations of MCFC joined with NGCC, and with PC let go control plants, the cathode gulf CO_2 fixation ranges from 4–5% to 11–12% separately. In exploratory examinations, Caprile et al. exhibited that a 125 kW pressurized MCFC could be worked with CO_2 focus as low as 4% in oxidant gas. Be that as it may, in another examination, Discepoli et al., utilizing a 55 cm^2 single cell, measured a limit centralization of 8% CO_2 underneath which the energy component encountered an extreme drop in execution if joined with low O_2 fixation (5%) and high oxidant use. At last, Di Giulio et al. effectively worked a 10x10 cm^2 single cell with as low CO_2 focus as 3%, in spite of the fact that the oxidant to fuel stream proportion was in the request of 30 to 1, which is significantly higher than what is conceivable in the lab scale MCFC utilized as a part of this work.

The constrained exploratory work which has been performed on MCFC in CCS conditions features the requirement for central research on as far as possible for working MCFC with CO_2 lean oxidant gas.

CHAPTER 13

APPLICATIONS OF FUEL CELL

CONTENTS

13.1. Power 152
13.2. Cogeneration 152
13.3. Automobiles 153
13.4. Buses 154
13.5. Forklifts 154
13.6. Motorcycles And Bicycles 154
13.7. Airplanes 155
13.8. Submarines 155
13.9. Portable Power Systems 155

13.1. POWER

Stationary energy units are utilized for business, mechanical and private essential and reinforcement control age. Energy components are exceptionally helpful as power sources in remote areas, for example, rocket, remote climate stations, extensive parks, interchanges focus, provincial areas including research stations, and in certain military applications. A power device framework running on hydrogen can be smaller and lightweight, and have no major moving parts. Since energy components have no moving parts and don't include burning, in perfect conditions they can accomplish up to 99.9999% unwavering quality. This likens to short of what one moment of downtime in a six-year term.

Since power device electrolyze frameworks don't store fuel in themselves, but instead depend on outside capacity units, they can be effectively connected in vast scale vitality stockpiling, country territories being one illustration. There are a wide range of sorts of stationary power modules so efficiencies fluctuate, yet most are in the vicinity of 40% and 60% vitality proficient. Be that as it may, when the energy unit's waste warmth is utilized to warm a working in a cogeneration framework this effectiveness can increment to 85%. This is fundamentally more productive than customary coal control plants, which are just around 33% vitality effective. Expecting generation at scale, power modules could spare 20–40% on vitality costs when utilized as a part of cogeneration frameworks. Energy components are additionally much cleaner than customary power age; a power module control plant utilizing petroleum gas as a hydrogen source would make short of what one ounce of contamination (other than CO_2) for each 1,000 kW·h created, contrasted with 25 pounds of toxins produced by ordinary burning frameworks. Energy units additionally deliver 97% less nitrogen oxide outflows than ordinary coal-let go control plants.

Energy units can be utilized with low-quality gas from landfills or wastewater treatment plants to produce power and lower methane discharges.

13.2. COGENERATION

Joined warmth and power (CHP) energy component frameworks, including Micro consolidated warmth and power (Micro CHP) frameworks are utilized to produce both power and warmth for homes, office building and manufacturing plants. The framework creates steady electric power (offering abundance control back to the matrix when it isn't expended), and in the mean time delivers hot air and water from the waste warmth. As the outcome CHP frameworks can possibly spare essential vitality as they can make utilization of waste warmth, which is for the most part dismissed by warm vitality transformation frameworks. A normal limit scope of home energy unit is 1–3 k_{Wel} 4–8 k_{Wth}. CHP frameworks connected to ingestion chillers utilize their waste warmth for refrigeration.

The waste warmth from power devices can be occupied amid the late spring straightforwardly into the ground giving further cooling while the waste warmth amid winter can be pumped specifically into the building. Co-age frameworks can achieve 85% productivity (40–60% electric + leftover portion as warm). Phosphoric-corrosive power modules (PAFC) involve the biggest section of existing CHP items worldwide and can give consolidated efficiencies near 90%. Liquid Carbonate (MCFC) and Solid Oxide Fuel Cells (SOFC) are likewise utilized for consolidated warmth and power age and have electrical vitality efficiencies around 60%. Drawbacks of co-age frameworks incorporate back sloping up and off rates, high cost and short lifetime. Additionally, their need a boiling water stockpiling tank to smooth out the warm warmth creation was a genuine inconvenience in the household commercial center where space in residential properties is at an extraordinary premium [107].

13.3. AUTOMOBILES

Exhibit power device vehicles have been created with "a driving scope of more than 400 km (250 mi) between refueling." They can be refueled in under 5 minutes. Division of Energy's Fuel Cell Technology Program guarantees that, starting at 2011, energy components accomplished 53–59% productivity at one-quarter control and 42–53% vehicle proficiency at full power, and a toughness of more than 120,000 km (75,000 mi) with under 10% debasement. In a Well-to-Wheels reenactment examination that "did not address the financial aspects and market limitations," General Motors and its accomplices evaluated that per mile voyaged, a power device electric vehicle running on compacted vaporous hydrogen created from flammable gas could use around 40% less vitality and produce 45% less nursery gasses than an interior burning vehicle. A lead design from the Department of Energy whose group is trying power device autos said that the potential interest is that "these are full-work vehicles without any confinements on range or refueling rate so they are an immediate swap for any vehicle. For example, on the off chance that you drive a full measured SUV and maneuver a vessel up into the mountains, you can do that with this innovation and you can't with current battery-just vehicles, which are more designed for city driving."

13.4. BUSES

As of August 2011, there were a sum of roughly 100 energy component transports conveyed the world over. These transports had aggregated more than 970,000 km (600,000 mi) of driving. Energy component transports have a 39–141% higher mileage than diesel transports and gaseous petrol transports. Power module transports have been conveyed the world over incorporating into Canada, San Francisco, United States, Germany, China, London, England, and Brazil [108].

13.5. FORKLIFTS

An energy component forklift (additionally called a power device lift truck) is a power device controlled modern forklift truck used to lift and transport materials. In 2013 there were more than 4,000 power module forklifts utilized as a part of material taking care of. The worldwide market is 1 million forklifts for each year. Power module armadas are worked by different organizations. Europe exhibited 30 power module forklifts with Hylift and extended it to 200 units, with numerous different tasks in line for future. Pike Research expressed in 2011 that energy component controlled forklifts will be the biggest driver of hydrogen fuel request by 2020.

Most organizations in Europe and the US don't utilize oil-fueled forklifts, as these vehicles work inside where discharges must be controlled and rather utilize electric forklifts. Energy unit controlled forklifts can give benefits over battery-fueled forklifts as they can work for an entire 8-hour move on a solitary tank of hydrogen and can be refueled in 3 minutes. Energy unit fueled forklifts can be utilized as a part of refrigerated stockrooms, as their execution isn't corrupted by bring down temperatures. The FC units are frequently planned as drop-in substitutions.

13.6. MOTORCYCLES AND BICYCLES

Maker of hydrogen-controlled power devices, Intelligent Energy (IE), created the primary working hydrogen run bike called the ENV (Emission Neutral Vehicle). The bike holds enough fuel to keep running for four hours, and to travel 160 km or 100 mi in a urban territory, at a best speed of 80 km/h or 50 mph.

13.7. AIRPLANES

Scientists and industry accomplices all through Europe directed exploratory flight trial of a kept an eye on plane controlled just by an energy component and lightweight batteries. The energy component demonstrator plane, as it was called, utilized a proton trade layer (PEM) energy component/lithium-particle battery half and half framework to control an electric engine, which was coupled to an ordinary propeller. The power module was a stack outline that enabled the energy component to be incorporated with the plane's streamlined surfaces.

Energy component controlled unmanned aeronautical vehicles (UAV) incorporate a Horizon power device UAV that set the record remove flown for a little UAV. The military is keen on this application in view of its low clamor, low warm signature and capacity to accomplish high height. Energy units are additionally being used to give helper control in flying machine, supplanting petroleum product generators that were beforehand used to begin the motors and power on load up electrical necessities.

13.8. SUBMARINES

As we all know the name the Type 212 submarines of the German and Italian naval forces utilize energy components to stay submerged for quite a long time without the need to surface. The U212A is a non-atomic submarine created by German maritime shipyard Howaldtswerke Deutsche Werft. The framework comprises of nine PEM power modules, giving between 30 kW and 50 kW each. The ship is quiet, giving it favorable position in the discovery of different submarines. A maritime paper has guessed about the likelihood of an atomic power device cross breed whereby the energy unit is utilized when quiet operations are required and afterward renewed from the Nuclear reactor and water too.

13.9. PORTABLE POWER SYSTEMS

Convenient power frameworks that utilization energy components can be utilized as a part of the relaxation division i.e., RVs, lodges, marine, the modern segment that is control for remote areas including gas and oil well destinations, correspondence towers, security, climate stations, and in the military segment.

PART III
SUPER CAPACITORS

CHAPTER 14

INTRODUCTION TO SUPER CAPACITORS

CONTENTS

14.1. Electrolytic Capacitors Before Super Capacitors 160
14.2. Emergence/Background 163
14.3. Overview 164
14.4. Construction And Working 165
14.5. Energy And Power 170
14.6. Devices And Technology Of Super Capacitors 170
14.7. Winding Technology 171

14.1. ELECTROLYTIC CAPACITORS BEFORE SUPER CAPACITORS

The capacitors of new generation are the electrolytic capacitors; they are of Ta, Al, and artistic electrolytic capacitors. Electrolytic capacitors utilize an electrolyte as conveyor between the dielectrics and a cathode. A run of the mill aluminum electrolytic capacitor incorporates an anode thwart and a cathode thwart handled by surface growth and additionally arrangement medications. More often than not, the dielectric film is created by anodizing high immaculateness Al thwart for high voltage applications in boric corrosive arrangements. The thickness of the dielectric film is identified with the working voltage of the aluminum electrolytic capacitor. In the wake of slicing to a particular size as indicated by the plan detail, an overlay made up of an anode thwart, a cathode thwart which is against the dielectric film of the anode thwart and a separator intervened between the anode and cathode foils, is twisted to give a component. The injury component does not have any electrical qualities of electrolytic capacitor yet until totally dunked in an electrolyte for driving and housed in a metallic sheathed bundle in round and hollow frame with a shut end preparing a releaser. Moreover, a fixing material made of versatile elastic is embedded into an open-end area of the sheathed bundle and the open-end segment of the sheathed bundle by drawing, whereby an aluminum electrolytic capacitor is constituted. Electrolytic aluminum capacitors are for the most part utilized as power supplies for autos, air ship, space vehicles, PCs, screens, motherboards of PCs and different gadgets [109]. There are two sorts of tantalum capacitors industrially accessible in the market; wet electrolytic capacitors which utilize sulfuric corrosive as the electrolyte and strong electrolytic capacitors which utilize MnO_2 as the strong electrolyte. Despite the fact that the capacitances got from both Ta and Al capacitors are the same, Ta capacitors are better than Al capacitors in temperature and recurrence attributes. For simple flag frameworks, Al capacitors deliver a present spike clamor which does not occur in Ta capacitors. At the end of the day, Ta capacitors are favored for circuits which require high strength attributes. The aggregate overall generation of Al electrolytic capacitors adds up to US$ 3.8 billion, 99% of which are of the wet sort. Dissimilar to Ta strong electrolytic capacitors, the strong electrolyte materials utilized are of natural starting point; poly pyrrole, a practical polymer and TCNQ (7,7,8,8 tetra cyanoquniodimethane) a natural semiconductor. Next, MnO_2 strong electrolyte material is shaped on the surface of that dielectric layer and over that a layer of poly pyrrole natural strong electrolyte material is framed by electrolytic blend. Following this, the positive and negative anodes are mounted to finish the electronic segment. Be that as it may, the capacitances of these electrolytic capacitors are in the range 0.1 to 10 F with a voltage profile of 25 to 50 V.

The historical backdrop of advancement of electrolytic capacitors which were mass created in the past and additionally today is exhibited by Niwa and Taketani. Numerous specialists endeavor to enhance the execution of these electrolytic capacitors by altering the terminal or electrolyte. For the most part, the increments in powerful surface zone (S) are accomplished by electrolytic carving of aluminum substrate before anodization, yet now it faces with the farthest point. It is additionally exceptionally hard to diminish d in light of the fact that the d esteem is generally chosen when the working voltages are chosen. Increment in might be a conceivable routine to shape composite dielectric layers by consolidating moderately substantial esteem mixes. Substitution of MnO by poly pyrrole strong electrolyte was accounted for to diminish electrostatic protection because of its higher conductivity; sweet-smelling sulfonate particles were utilized as charge repaying dopant particles. A tantalum capacitor with Ta metal as anode, poly pyrrole as cathode and Ta_2O_5 dielectric layer was additionally announced. In the Al strong electrolytic capacitors, polyaniline doped with inorganic and natural acids was likewise considered as counter cathode. In yet another work, Al strong electrolytic capacitor with scratched Al thwart as anode, polyaniline or poly pyrrole as cathode and Al_2O_3 as dielectric was produced. Ethylene carbonate based natural electrolytes and butyrolactone based electrolytes have been attempted as working electrolytes in Al electrolytic capacitors. Masuda et al. have gotten high capacitance by electrochemically anodizing quickly extinguishing Al-Ti compound thwart. Numerous scientists have attempted the other blend of amalgams, for example, Al-Zr, Al-Si, Al-Ti, Al-Nb, and Al-Ta composite oxide films. Composite oxide movies of Al_2O_3-(Ba 0.5 Sr 0.5 TiO_3) and $Al_2O_3Bi_4Ti_3O_{12}$ on low-voltage carved aluminum thwart were additionally contemplated. Nb-Ta-Al for Ta electrolytic capacitors was additionally attempted as anode material. A clay capacitor is a capacitor developed of substituting layers of metal and earthenware, with the artistic material going about as the dielectric. Multilayer fired capacitors (MLCs) ordinarily comprise of about 100 substitute layers of cathode and dielectric earthenware production sandwiched between two clay cover layers. They are created by screen-printing of anode layers on dielectric layers and co-sintering of the overlay. Ordinarily, Ag-Pd is utilized as the anode material and $BaTiO_3$ is utilized as the dielectric earthenware production. From

2000 onwards, the MLCs showcase has been developing in pace with the exponential improvement of correspondences. They are delivered in the capacitance scope of 10 F (ordinarily the scope of Ta and Al electrolytic capacitors); they are very helpful in high recurrence applications. Truly, an earthenware capacitor is a two-terminal non-polar gadget. The established clay capacitor is the plate capacitor. This gadget originates before the transistor and was utilized broadly in vacuum-tube hardware (e.g., radio recipients) from c.a. 1930 through the 1950s and in discrete transistor hardware from the 1950s through the 1980s. Starting at 2007, fired circle capacitors are in far reaching use in electronic gear, giving high limit and little size at low value contrasted with alternate sorts. The other earthenware materials that have been recognized and utilized are $CaZrO_3$, $MgTiO_3$, $SrTiO_3$ and so on. An average 10 F MLC is a chip of size (3.2 x 1.6 x 1.5 mm). Mn, Ca, Pd, Ag, and so on are a portion of the other inner terminals utilized. Direct dielectrics and antiferroelectric-based o strontium titanite have been created for high voltage circle capacitors. These are appropriate for MLCs with more slender layers in view of their high coercive fields. A standout amongst the most basic material handling parameters is the level of homogeneous blending of added substance in the slurry. The fastener conveyance in the green artistic sheet, the level of surface harshness, fine size nickel powder, arrangement of green sheet, anode affidavit advertisement sheet stacking and so on assume a significant part in the process innovation. Any of these actualities if misused would bring about the disappointment of the gadget. For example, giving a roughness of 5 m thick green sheet to 0.5 m is compulsory so a smooth contact surface with the internal nickel anode can be built up. This is an imperative factor in maintaining a strategic distance from the centralization of electric documented at severities, where the charge outflow from the cathode is quickened, bringing about short disappointment. Customary sheet/printing technique has a specialized cutoff of creating a thickness around 1 m dielectric; keeping in mind the end goal to diminish the thickness further, thin film advancements like CVD, sputtering, plasma-shower, and so on must be utilized. Alternate sorts of capacitors are film capacitors which utilize thin polyester film and polypropylene film as dielectrics and meta-coated capacitors which consolidate terminal plates made of film vacuum vanished with metal, for example, Al. Movies can be of polyester, polypropylene or polycarbonate make. Likewise, capacitors are indicated relying upon the dielectric utilized, for example, polyester film capacitor, polypropylene capacitor, mica capacitor, metallized polyester film capacitor, etc.

14.2. EMERGENCE/BACKGROUND

Environmental change and the constrained accessibility of non-renewable energy sources have incredibly influenced the world's economy and nature. With a quickly developing business sector for convenient electronic gadgets and the improvement of cross breed electric vehicles, there has been a regularly expanding and critical interest for ecologically inviting high-control vitality assets. The creation of electrochemical vitality is under genuine thought as an option vitality/control source, as long as the utilization of this vitality is intended to be more economical and all the more ecologically well disposed. Frameworks for electrochemical vitality stockpiling and change incorporate batteries, energy components and electrochemical capacitors (ECs). ECs are referred to by various names, for example, ultra capacitors, EDLC, or super capacitors. These names were developed by various produces of the ECs. The exchange name of the main business gadget made by Nippon Electric Company (NEC) was Super capacitors, however Pinnacle Research Institute (PRI) called the ECs ultra capacitors. Whatever the exchange name of ECs, they all allude to a capacitor, which stores electrical vitality in the interface between an electrolyte and a strong anode. Due to the low capacitance estimations of the electrostatic capacitors, they are constrained to low power applications or at most here and now memory go down provisions. A substantial capacitance esteem and a high working cell voltage are required for a super capacitor to have great execution. Henceforth, the advancement of both novel anode materials with expanded capacitance, for example, grapheme-based materials, and electrolytes with more extensive potential windows, for example, ionic fluid electrolytes or natural electrolytes, is required to improve the general execution of the super capacitor.

ECs are now and then called Super capacitors, ultra capacitors, or cross breed capacitors. The term ultra capacitor or super capacitor is typically used to depict a vitality stockpiling gadget in light of the charge stockpiling in the electrical twofold layer (EDL) of a high-surface region carbon in watery electrolytes. The market for these gadgets utilized for memory assurance in electronic hardware is about \$150–\$200 million every year. New potential applications for ECs incorporate the compact electronic gadget advertise, the power quality market, due especially to circulated age, and low-discharge crossover autos, transports, and trucks. There is no regularly acknowledged classification for ECs aside from that the meanings of anode, cathode, and so on, persist from battery and power device utilization, for example, the anode as the negative terminal and cathode as the positive terminal. All in all, ultra capacitors have alluded to capacitors with two high-surface region carbon terminals for the anode and cathode. This course of action where the two cathodes have a similar design will be alluded to as a symmetric capacitor. The term super capacitor has additionally been utilized to allude to the symmetric blend of two carbon anodes that are catalyzed with ruthenium dioxide (RuO_2). The RuO_2 presents a redox couple between two valence conditions of ruthenium, to bring about higher capacitance for the carbon terminals, however with a punishment in slower time steady to react to a heartbeat request. A moment kind of EC joins a battery or redox terminal with a carbon anode, for example, nickel hydroxide cathode with a carbon anode. These super capacitor or cross breed capacitors will be alluded to as an uneven EC [110].

14.3. OVERVIEW

The third segment concentrates on electrochemical twofold layer capacitors or Super capacitors, taking a gander at carbon based frameworks, which have the most modern importance. Here they introduce a definite examination of cathode enhancement for various types of carbon, for example, initiated carbon, carbide subsidiaries, aerogels, nanotubes, and graphene, with different electrolyte sorts as far as pore measure appropriation and the impact of practical gatherings. The discourse at that point takes after onto a scope of electrolyte sorts: watery, non-fluid, and ionic fluids. Perceptions propose that in the long run promote development in specific surface territory of terminals through diminishing pore sizes won't bring about enhanced super capacitor vitality thickness due to the influence of steric impacts. Accordingly, there is an expanded enthusiasm for pseudo capacitor cathode materials which incorporates metal oxides, for example, iridium (IV) oxide (IrO_2), manganese(IV) oxide (MnO_2) and ruthenium (IV) oxide (RuO_2), directing polymers and monomer redox frameworks. This content gives an exhaustive diagram of these terminal materials, difficulties and prospects, of which a present pattern is seen to be the formation of composite cathodes of metal oxides and directing polymers to alleviate the conductivity issues of the metal oxides and accomplish higher mass loadings for more down to earth gadgets. The definite discourse of carbon based super capacitor terminals and pseudo capacitor anodes at that point leads onto awry Super capacitors which joins the two together to give a higher vitality thickness because of a more extensive working voltage window. Reflecting on the greater part of this, the segment closes with an examination of financially accessible Super capacitors and the possibilities of the gadget [111].

Capacitors store electrical charge. Since the charge is put away physically, with no concoction or stage changes occurring, the procedure is exceptionally reversible and the release charge cycle can be rehashed again and again, for all intents and purposes unbounded. Electrochemical capacitors (ECs), differently alluded to by producers in limited time writing as "super capacitors" or "ultra capacitors," store electrical charge in an electric twofold layer at the interface between a high-surface-region carbon terminal and a fluid electrolyte. Therefore, they are likewise legitimately alluded to as electric twofold layer capacitors.

Super capacitors don't contain a dielectric square. The electrical twofold layer are framed in the electrolyte encompassing the particles, prompting viable detachment of charge on the request of nanometer scale. The territory of the electrical twofold layer relies upon the surface region of the particles. High capacitances result from the viable measured bundles. In an electrical twofold layer, each layer without anyone else's input is very conductive, yet the material science at the interface where the layers are viably in contact implies that no huge current can stream between the layers. Be that as it may, the twofold layer can withstand just a low voltage. Since the capacitance of these gadgets is corresponding to the dynamic terminal territory, expanding the anode surface zone will build the capacitance, thus expanding the measure of vitality that can be put away. The electric twofold layer capacitor (EDLC) is perfect for vitality stockpiling that experiences visit charge and release cycles at high present and brief span. Farad is a unit of capacitance named after the English physicist Michael Faraday. One farad stores one coulomb of electrical charge while applying one volt.

14.4. CONSTRUCTION AND WORKING

A straightforward EC can be built by embedding two conductors in a measuring glass containing an electrolyte, for instance, two carbon bars in salt water. At first there is no quantifiable voltage between the two poles, yet when the switch is shut and current is caused to spill out of one bar to the next by a battery, charge partition is normally made at every fluid strong interface. This adequately makes two capacitors that are arrangement associated by the electrolyte. Voltage perseveres after the switch is opened vitality has been put away. In this state, solvated particles in the electrolyte are pulled in to the strong surface by an equivalent yet inverse charge in the strong. These two parallel districts of charge shape the wellspring of the expression "twofold layer." Charge partition is measured in atomic measurements (i.e., couple of angstroms), and the surface region is measured in a huge number of square meters per gram of terminal material, making 5 kF capacitors that can be hand-held. The very component of an electrochemical capacitor that makes such high capacitances conceivable, in particular the profoundly permeable high surface territory terminals is additionally the purpose behind the moderately moderate reaction of these gadgets contrasted with customary capacitors. Electrical associations with the put away charge are made through the strong carbon encompassing the pore and through the electrolyte from the mouth of the pore, electrolyte conductivity being considerably less than carbon conductivity. Charge put away close to the pore mouth is open through a short way with little electrolyte protection. Interestingly, charge put away more profound inside the pore must cross a more extended electrolyte way with a fundamentally higher arrangement protection. Hence, the general reaction can be spoken to by a numerous time-steady proportional circuit show. Independent of this conduct, the reaction time of an electrochemical capacitor in both charge and release operation is to a great degree short, around 1 second, when contrasted with batteries (minutes to many minutes). The working voltage of an electrochemical capacitor is constrained by the breakdown capability of the electrolyte (commonly 1 to 3 V for every cell) and is for the most part much lower than that of regular electrostatic and electrolytic capacitors. In numerous useful applications, in this way, electrochemical capacitor cells must be arrangement associated, like batteries, to meet working voltage necessities. The key distinction amongst batteries and electrochemical capacitors is that the previous store vitality in the majority of compound reactants equipped for producing charge, while the last store vitality specifically as surface charge. Battery release rate and thusly control execution is then restricted by the response energy and in addition the mass transport, while such confinements don't have any significant bearing to electrochemical capacitors built with two enacted carbon anodes, consequently permitting uncommonly high-power ability amid both release and charge. Most batteries display a moderately consistent working voltage in view of the thermodynamics of the battery reactants; as an outcome, it is regularly hard to gauge their condition of-charge (SOC) correctly. Then again, for a capacitor, its working voltage changes directly with time amid consistent current operation so that the SOC can be precisely pinpointed. Besides, the exceedingly reversible electrostatic charge stockpiling system in ECs does not prompt any volume change like saw in batteries with electrochemical changes of dynamic masses. This volume change restrains the cyclability of batteries for the most part to a few hundred cycles while ECs have shown from several thousands to a huge number of full charge/ release cycles. Close to twofold layer capacitors, there is a class of vitality stockpiling materials that experience electron exchange responses yet carry on in a capacitive way. These materials store the vitality utilizing profoundly reversible surface redox (faradaic) responses notwithstanding the electric twofold layer stockpiling, in this way characterizing pseudocapacitive storage. Materials that display pseudo capacitance go from leading polymers to an assortment of progress metal oxides.

The most well-known is the RuO_2 pseudo capacitor that can reach over 800 F/g, high volumetric and power densities, and shows incredible cycle life. On the drawback, RuO_2 is a respectable metal oxide and in that capacity dreadfully costly for most business applications. Endeavors to grow more viable pseudocapacitive materials are very dynamic right now. The most punctual electrochemical capacitors presented 30 years prior were symmetric outlines (two indistinguishable cathodes) in fluid electrolyte, H_2SO_4 or KOH, which restricted working cell voltage to 1.2 V/cell and ostensible cell rating to about 0.9 V. In the second era of electrochemical capacitors, the utilization of natural electrolyte that is ordinarily an ammonium salt disintegrated in a natural dissolvable, for example, propylene carbonate or acetonitrile is prompted an expansion of the evaluated cell voltage from around 0.9 V/cell to 2.3–2.7 V/cell. Winding injury or kaleidoscopic plate development electrochemical capacitors utilizing a natural electrolyte are the most well-known sort today. The latest electrochemical capacitor plans are awry and included two capacitors in arrangement, one capacitor-like and the other a pseudo capacitor or battery like, with fluctuating terminal limit proportions, contingent upon the application. The capacitor anode is indistinguishable to those utilized as a part of symmetric electrochemical capacitors. Conversely, the battery like or pseudo capacitor terminal depends on very reversible redox (electron charge exchange) responses. In this outline, the limit of the battery-like cathode is by and large commonly more noteworthy than the limit of the twofold layer capacitor anode, which is the reason for the name "lopsided." In looking at the two plans, both with precisely the same twofold layer charge stockpiling terminal and electrolyte, the topsy-turvy configuration gives precisely double the capacitance of the symmetric outline. This happens in light of the fact that the electron-exchange terminal's potential is basically settled, with just the carbon anode potential changing with charge. Additionally, the working voltage of the hybrid configuration is bigger, because of the two terminals having distinctive rest possibilities. Both of these elements add to a higher vitality thickness than can be accomplished with a symmetric outline. A few novel topsy-turvy electrochemical capacitor plans are under development 10–12 utilizing a lithium particle intercalation anode in a natural electrolyte at 3.8 V 10 or a carbon terminal with a lead dioxide battery-like cathode utilizing H_2SO_4 as the electrolyte, working at 2.1 V with the capability of being low cost. Each of these outlines can give high cycle-life because of the cathode limit asymmetry. Extensive research accentuation is being given to uneven electrochemical capacitors today on account of the extremely alluring execution highlights.

Electrochemical capacitors are close cousins to batteries. The basic circuit demonstrated shows their essential operation (Figure 14.1).

Figure 14.1: Simple circuit. (Source: *https://www.google.com/search?client=firefox-b-ab&biw=1366&bih=695&tbm=isch&sa=1&ei=iPFBWo-QEoiAU-b1p_AE&q=simple+RC+circuit&oq=simple+RC+circuit&gs_l=psy-ab.3.0l2j0i24k1l3.65884.72660.0.73091.32.15.0.5.5.0.550.2120.3–3j0j2.6.0....0...1c.1.64.psy-ab..21.9.1843. 0..0i67k1.449.92mv_hJvUw0*).

Here, C_a and C_c are the twofold layer capacitances of the anode and cathode, individually. Ri is the inward protection of the cell. For capacitors in series

$$\frac{1}{C} = \frac{1}{C_a} + \frac{1}{C_c} \tag{1}$$

If C_a and C_c, as would be expected for an ultra capacitor, then

$$C = \frac{1}{2} C_a \tag{2}$$

High surface territory carbon is the material of decision, as it joins an expansive surface region wetted by the electrolyte, high electronic conductivity, and substance and electrochemical strong qualities with minimal effort. The capacitance of these gadgets can be requests of greatness bigger than those of customary dry and electrolytic capacitors. The voltage for electrochemical capacitors with fluid electrolytes is 1 V, restricted by the voltage strength of the electrolyte. By changing to a natural based electrolyte, voltages of up to 2.7 V can be found practically speaking. Be that as it may, the natural electrolytes have brought down twofold layer capacitance and poorer conductivity. Since the vitality stockpiling is given by vitality ½ QV_2, the higher voltage allowed by a natural electrolyte fundamentally expands the vitality stockpiling ability of the EC. Since the resistivity is 100 times bigger than for fluid electrolytes, the time steady for reaction to an expansive heartbeat is slower for the non-aqueous electrolyte based ECs. The charge-release of a symmetric EC made out of two carbon anodes with around a similar mass drenched in a fluid or non-aqueous electrolyte. With zero connected charge Q, the two anodes of a cell are at a similar voltage. The capability of the terminals increments in inverse ways amid charge, as every ha around a similar capacitance. Greatest cell working voltage is achieved when one of the cathodes achieves the soundness furthest reaches of the electrolyte. The hybrid sort of EC joins a battery anode as one of the terminals. The battery anode has a capacitance related with the redox battery response of 10 times the capacitance of the electrical twofold layer. For example, if the nickel battery cathode is substituted for the cathode in a symmetric capacitor, for instance, NiOOH for carbon, at that point substituting $C_c = 10\ C_a$ in eq 14.3, the capacitance of the EC is basically doubled.

$$\frac{1}{C} = \frac{1}{C_a} + \frac{1}{10C_a} \tag{3}$$

$$C \approx C_a \tag{4}$$

This uneven sort of EC is regularly named a "cross breed" capacitor. The regular release bend for this mixture EC. Since the battery anode has a limit of 3–10 times that of the twofold layer terminal, it stays at an invariant voltage amid charge and release. Thus, the release voltage of the half and half capacitor falls more gradually than that of the carbon-carbon EDL capacitor. Now and again, the energy of the redox charge release responses can continue nearly as fast and reversibly as EDL charging. Thin film redox terminals, in view of the lithium intercalation/inclusion standard, for example, $Li_4Ti_5O_{12}$, show high reversibility and quick energy. The RuO_2 materials saved on carbon indicate "pseudo-capacitive" charge release conduct as do polymeric materials, for example, polyaniline, poly pyrrole, and poly di amino anthraquinone (DAAQ). These have effortless energy and have demonstrated high capacitance and long life. The inclusion of anions and cations into their structure can yield capacitances of up to 200 $\mu F/cm^2$ and, besides, they can be effortlessly created as thin films.

14.5. ENERGY AND POWER

The maximum energy E (J) stored in a super capacitor is given by equation

$$E = \frac{1}{2} CU^2 \quad (5)$$

where C is the particular capacitance (F), U is the greatest cell voltage (V). The capacitance depends basically on the cathode materials utilized. The voltage is restricted by the steadiness potential window of the electrolyte. For the most part, the greatest voltage is 0.6–0.8 V in fluid electrolytes and 2.5–2.7 V in natural medium. A further development of the voltage range can be acknowledged by utilizing ionic fluids as it will be talked about later. The most extreme power (P) of a super capacitor is figured by equation

$$P = \frac{U^2}{4R_s} \quad (6)$$

where U is the most extreme cell voltage (V), RS the proportionate arrangement protection (ESR in Ω) and P the greatest power (W). The elements which mostly add to the general arrangement protection of Super capacitors are the conductivity of the anode material, the contact protection between the terminal material and the present authority, the electrolyte protection, the ionic dissemination protection because of the developments in micro pores and the ionic protection caused by the separator. Because of the electrostatic charge stockpiling system, the arrangement protection does exclude any charge exchange protection commitment related with electron trade, as watched for redox responses. In this way, the arrangement protection is lower than that of batteries at cell level, clarifying the higher power thickness of Super capacitors contrasted with batteries. The ESR is lessened by including a percolator which enhances the conductivity of the terminal and by enhancing the electrical contact between the present gatherers and the dynamic materials.

14.6. DEVICES AND TECHNOLOGY OF SUPER CAPACITORS

Two development sorts of super capacitor gadgets are by and large utilized: the monopolar cell outline and the bipolar cell plan.

14.6.1. Monopolar Cell

The monopolar gadget indicates high capacitance because of a high surface territory. The gadgets are amassed by winding or stacking in parallel indistinguishable terminals, separator and authority foils. The voltage of the gadget is indistinguishable to that of the single cell. It is likewise conceivable to create lopsided gadgets with one super capacitor terminal and one battery cathode, for example, NiOx or MnO_2. The thought is to twofold the capacitance of the gadget on the grounds that lone the super capacitor-cathode decides the capacitance now. Also, the cell voltage is higher because of the redox capability of the battery cathode. As leeway, the voltage of such Super capacitors amid release diminishes just marginally. Vitality densities of up to 20 Wh/kg and power densities of up to 5 kW/kg are conceivable. Another kind of lopsided gadget is the crossover capacitor by Evans. He utilizes RuO_2 for a high capacitance cathode and a Ta electrolytic capacitor terminal. This gadget was made for military applications and opposed a few thousand grams increasing speed. A high profitability is acquired by winding the distinctive monopolar cell parts with an exact control of the considerable number of parameters.

14.6.2. Bipolar Cell

The bipolar gadgets permit a high voltage, low protection and low capacitance [16]. Such super capacitors are developed by stacking a few single cells in arrangement into one case. The fixing of such gadgets is exceptionally refined, as the electrolyte must be kept from reaching one cell with another.

14.7. WINDING TECHNOLOGY

The bipolar gadgets permit a high voltage, low protection and low capacitance. Such Super capacitors are developed by stacking a few single cells in arrangement into one case. The fixing of such gadgets is extremely advanced, as the electrolyte must be kept from reaching one cell with another ESR.

CHAPTER 15

PROPERTIES OF SUPER CAPACITORS

CONTENTS

15.1. Equivalent Circuit Of Super Capacitors 174
15.2. Plot Of Ragone 176
15.3. Power Availability 178
15.4. Energy Availability 179
15.5. Double Layer Capacitors (DLC) 180
15.6. Performance 192
15.7. Distinction Between Super Capacitor And Capacitor 194

15.1. EQUIVALENT CIRCUIT OF SUPER CAPACITORS

The capacitor might be modelized by the electrical circuit. The equal arrangement protection ESR confines the current and is in charge of the electrical misfortunes. To get high power, it is totally important to have a low arrangement protection. The parallel protection Rp is in charge of the capacitor self-release time. Its esteem must be as high as conceivable to restrict the spillage current. The time consistent τ of the self-release is τ = Rp C.

Proportional circuits are for the most part compelled by an arrangement of limit conditions, whose object is to confine the quantity of applicant answers for the current issue. Shockingly, unique limit conditions are regularly accepted in various orders, making cross-disciplinary correlations hazardous. A particularly vexing circumstance emerges when contrasting outcomes got by electrical designers and electrochemists. Among electrical architects, the expression "identical circuit" by and large alludes to a basic arrangement mix of two parts, (for example, a resistor and a capacitor) that copies the reaction of are al framework at a solitary determined recurrence. (This utilization is pushed by the American National Standards Institute (ANSI), for instance.) Such a circuit is appeared in Figure 15.1. For this situation, there is no necessity that the identical circuit should coordinate the sufficiency and stage move of the genuine framework at some other recurrence. Among electrochemists, the expression "comparable circuit" has an altogether different significance. It alludes to a bigger system that copies the reaction of a genuine framework at different determined frequencies. Despite the fact that there is no confinement on the measure of such a system, electrochemists constantly apply Occam's razor thus don't embed a larger number of segments into the system than there are physical marvels to be clarified. By the by, they for the most part think about an expansive variety of systems as competitor arrangements and after that select the "best" system as the one whose segment esteems limit the entirety of the squared residuals between the hypothetical impedance work and the exploratory impedance work. By differentiate, no such choice process is connected in the building case. The presence of these two unique methods of insight of equal circuit demonstrating implies that distributed estimations of "capacitance" must be translated with alert. To represent this point, consider an arbitrarily designed electrical system having a genuine physical capacitor implanted some place inside it. An express equation for the capacitance of this capacitor is given by Gauss' law, which expresses that the put away charge Q is equivalent to the indispensable of the ordinary segment of the neighborhood electric field E (a vector) over the zone An, increased by the nearby outright permittivity ε.

$$Q = \oint \varepsilon \vec{E}.d\vec{A} \tag{1}$$

As needs be, the proportion $Q/V=C_{real}$ is the genuine capacitance of intrigue, which for a parallel plate capacitor of thickness d takes the form [112]

$$C_{real} = \frac{\varepsilon A}{d} = \frac{\varepsilon_r \varepsilon_0 A}{d} \tag{2}$$

where ε_0 is the supreme permittivity of the vacuum and εr is the neighborhood relative permittivity (dielectric consistent) of the medium inside the capacitor. Be that as it may, just an electrochemical proportional circuit could recognize C_{real}. A designing identical circuit would neglect to do as such in light of the fact that its "comparable arrangement capacitance" is an anecdotal capacitance whose sole reason for existing is to imitate the plentifulness and stage reaction of the aggregate framework. In this manner, the identical arrangement capacitance has no straightforward physical correlative. To be sure, for the equal arrangement circuit, it is given by the equation

$$C_{ESC}\left(w\right) = \frac{1}{jwImZ\left(w\right)} \tag{3}$$

where j is the square foundation of short one, ω is the precise recurrence, and Im Z(ω) is the nonexistent segment of the intricate impedance of the aggregate framework. Note that, dissimilar to the genuine physical capacitance, the identical arrangement capacitance got from this recipe might be (generally is) recurrence subordinate. This unphysical result emerges in light of the fact that the proportional arrangement capacitance is constrained to take after the reaction of the framework all in all (as shown in Eq. (3), not only the reaction of the segregated capacitor as in Eq. (2). Consequently,

$$C_{ESC}\left(w\right) \neq \frac{\varepsilon A}{d} \tag{4}$$

Because of these contemplations, writing plots of capacitance versus "surface territory" are aimless unless the writers have characterized precisely what they mean by capacitance and ideally showed that this same capacitance isn't recurrence subordinate. It is clear from the above discourse that two distinctive ways to deal with comparable circuits are in like manner utilize, and these create two unique estimations of capacitance for a similar genuine framework. On one hand, there is a two-segment ("designing") proportional circuit which creates a recurrence subordinate equal arrangement capacitance whose properties may some of the time abuse presence of mind. Then again, there is a numerous part ("electrochemical") proportionate circuit which produces a recurrence free genuine capacitance (C_{real}) whose properties are very much acted. It involves lament that, in the logical writing, these methodologies are not generally plainly recognized. An especially intolerable illustration is the capacitance inferred by measuring the standard hysteresis of voltammograms, which is almost dependably a proportional arrangement capacitance, despite the fact that this reality is once in a while (if at any point) specified to the reader. A comparable surrender at applies to the instance of capacitance estimations acquired by connect techniques; these normally produce estimations of proportional arrangement capacitance as opposed to genuine physical capacitance.

15.2. PLOT OF RAGONE

The Ragone plot is a diagram, which gives the connection between the vitality thickness and the power thickness of a capacity framework. It is utilized to think about the exhibitions of the diverse stockpiling sorts like batteries, Super capacitors and regular capacitors.

The parameter of the Ragone plot is really the heap protection RL. It means that the plot demonstrates the vitality and the power, which can be disseminated for the diverse estimations of the heap protection. To rearrange the relations, we present the parameter α characterized by the connection τ = α τ0, RL + RS= α RS. In the event that RL is consistent, the underlying pinnacle control at time t=0 might be composed as

$$P_{ab} = \frac{U_0^2(\alpha - 1)}{\alpha^2 / R_s} \tag{5}$$

The maximum peak power $P_{max,0}$ is available at time t = 0 when $R_S = R_L$, and it is equal to

$$P_{max,o} = \frac{E_{max}}{2t_0} = U_0^2 / 4R_s \tag{6}$$

Only a part of the stored energy is available for the load. The amount is equal to the load power integration from t = 0 to t = ∞.

$$E_a = E_{max}\left(a-1\right)/a \quad (7)$$

The Ragone plot coordinates for a given RL are consequently

$$E_a^{'} = \frac{E_a}{M} = \frac{E_{max}\left(a-1\right)}{a/M} \quad (8)$$

$$P_{a,o}^{'} = \frac{P_{a,o}}{M} = \frac{2E_{max}\left(a-1\right)}{\frac{\frac{a^2}{t_0}}{M}} \quad (9)$$

$$= \frac{4P_{max,o}\left(a-1\right)}{a^2/M} \quad (10)$$

The numerical estimations of these directions are normally given per unit of segment weight.

Super capacitors have drawn incredible consideration in view of their high charge-release rate, long cycle life, remarkable power thickness, and no short out issue that are of worries with current battery or energy component techniques. Otherwise called ultra capacitors or electrochemical capacitors, Super capacitors store vitality with electric twofold layer (EDL) capacitance (particle adsorption) or pseudo capacitance (surface redox response). Despite the fact that pseudo capacitors utilizing directing polymers or metal oxides as anode materials can have a high faradaic capacitive execution, they can't keep up it after long cycling. Then again, EDL capacitors with permeable carbon terminals can be charged and released upwards of one million cycles without execution degradation. Besides, the particle transportation in EDL capacitors is speedier than the redox response in pseudo capacitors, prompting a high charge-release rate and power thickness in EDL capacitors. As of now the significant downside of EDL capacitors is the low vitality thickness by and large in the scope of 3–5 Wh/kg, which is one to two requests of size beneath marketed lithium-particle batteries (100–275 Wh/kg). Super capacitors with expanded vitality stockpiling yet little forfeit of energy thickness are fundamentally required for reasonable applications, for example, module cross breed electric vehicles, wind turbine vitality storage, regenerative braking etc..

To enhance the vitality thickness, a permeable terminal material with profoundly created surface zone and high electric conductivity is wanted. Carbon materials, because of their ease, different micro texture and process ability, are more appealing than different materials. Actuated carbons, carbon aerogels, and carbon nanotubes have been generally used. Initiated carbons are the most widely recognized materials in EDL Super capacitors these days. The detailed particular surface territory of initiated carbon ranges from 500 to 2000 m^2/g, however just a small amount of it can add to vitality stockpiling. This is on account of the actuated carbon materials have poor consistency in pore sizes extending from micro pores of about 0.3 nm to macro pores. In such materials, the micro pores are difficult to reach to electrolyte solutions and macro pores result in a low surface volume proportion, both of which prompt a low particular capacitance. Carbon nanotubes have a direct surface region and in addition a decent conductivity. Super capacitors in view of carbon nanotubes can't show great capacitive execution without including a pseudocapacitive component, and the assembling troubles in addition to cost additionally confine their probability of usage in real vitality stockpiling gadgets.

15.3. POWER AVAILABILITY

At the point when the heap protection is littler than the inward protection, the present will be higher, however the voltage drop on the heap is emphatically lessened. The result is that a littler power might be conveyed to the heap. On the correct side of the Ragone plot that is shown in Figure 15.1, the bend is hindered on a state of greatest power thickness $P'_{max,0}$ relating to a heap protection equivalent to the capacitor inward protection. On the left side, in the "low" power run the heap protection is substantially greater than the capacitor inner protection.

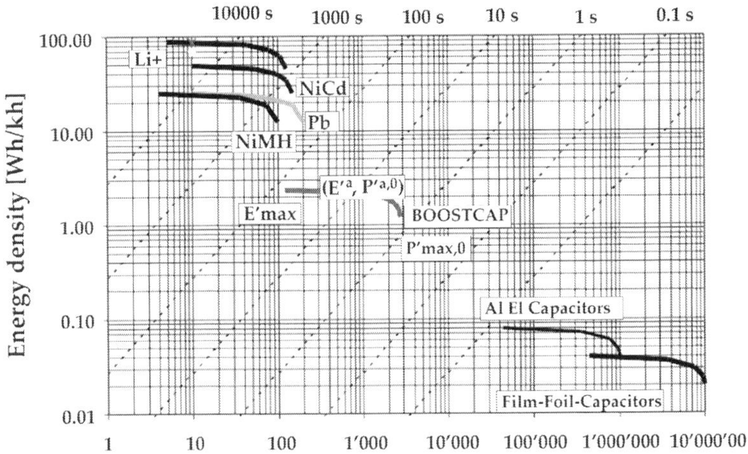

Figure 15.1: Energy density vs. power density. (Source: Schneuwly, A. and Gallay, R. (2000). montena components SA, 1728 Rossens, Switzerland. Properties and applications of super capacitors from the state-of-the-art to future trends).

15.4. ENERGY AVAILABILITY

The main perception, which might be done, is that at the most extreme power rate, just 50% of the put away vitality is accessible. The second half is dispersed in the capacitor inner protection. In this powerful condition, the current is critical and the misfortunes in the capacitor are relative to the interior protection time the square of the current. The second perception is that in the "low" power condition all the vitality E' max is accessible for the heap. The misfortunes are little in light of the fact that for a similar capacitor inner protection, the current is little. All the above comments are legitimate for the essential and auxiliary batteries. The misfortunes inside the capacitor are

$$E_{p\alpha} = \frac{E_{max}}{a} P_{p\alpha b} = \frac{2E_{max}}{a^2 / t_0} \quad (11)$$

The efficiency η which is the ratio between the energy used in the load to the total energy content, is equal to η = 1

$$\alpha = \frac{RL}{RL + RS} \quad (12)$$

15.5. DOUBLE LAYER CAPACITORS (DLC)

Electric/electrochemical twofold layer capacitor (EDLC) is a one of a kind electrical stockpiling gadget, which can store significantly more vitality than regular capacitors and offer considerably higher power thickness than batteries. EDLCs top off the hole between the batteries and the ordinary capacitor, permitting applications for different power and vitality prerequisites i.e., move down power hotspots for electronic gadgets, stack leveling, motor begin or quickening for half breed vehicles and power stockpiling created from sunlight based or wind vitality. EDLC chips away at the rule of twofold layer capacitance at the terminal/electrolyte interface where electric charges are gathered on the cathode surfaces and particles of inverse charge are orchestrated on the electrolyte side. There are two principle sorts of twofold layer capacitors as characterized by the charge stockpiling component:

(1) electrical twofold layer capacitor.

(2) electrochemical twofold layer capacitor or super/pseudo-capacitor.

An EDLC stores vitality in the twofold layer at the cathode/electrolyte interface, while the super capacitor maintains a Faradic response between the terminal and the electrolyte in a reasonable potential window. Subsequently the terminal material utilized for the development of the cell for the previous is essentially carbon material while for the last mentioned, the cathode material comprises of either progress metal oxides or blends of carbon and metal oxides or polymers. The electrolytes can be either fluid or non-watery relying upon the method of development of EDLC cell. There are two general bearings of intrigue. One is the long-haul objective of the advancement of electrical impetus for vehicles, and the other is the quick development of compact electronic gadgets that require control sources with greatest vitality content and the most minimal conceivable size and weight.

15.5.1. Operating Principal of Double Layer Capacitors

An ES is a charge-stockpiling gadget like batteries in plan and assembling. As appeared in Figure 15.2, electrochemical capacitors (super capacitors) comprise of two anodes isolated by a particle penetrable layer (separator), and an electrolyte interfacing the two terminals electrically. By applying a voltage to the capacitor, an electric twofold layer is framed at the two cathodes, which has a positive or negative layer of particles saved in an identical representation on the contrary terminal. The standards of a solitary cell twofold layer capacitor and a delineation of the potential drop at the anode/electrolyte interface are appeared in Figure 15.2 [114].

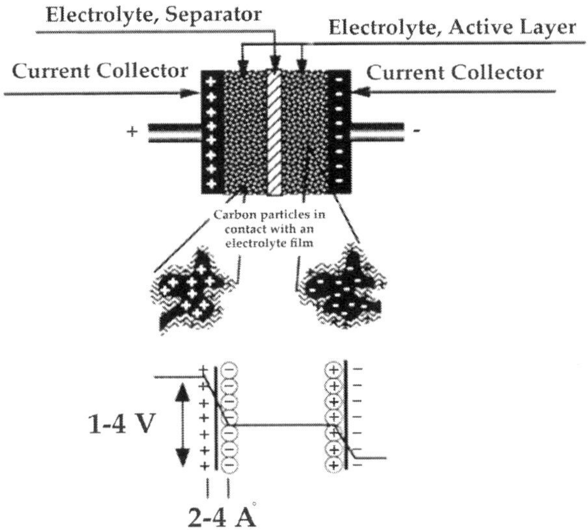

Figure 15.2: Interface of electrode and electrolyte. (Source: Elaheh Kowsari, E. (2018). *High-Performance super capacitors based on ionic liquids and a graphene nanostructure*, 10(1)).

The two electrodes form a series circuit of two individual capacitors C_1 and C_2. The total capacitance C_{total} is given by

$$C_{total} = \frac{C_1.C_2}{C_1 + C_2} \tag{13}$$

Super capacitors may have either symmetric or unbalanced terminals. Symmetry suggests that the two anodes have a similar capacitance esteem. In the event that $C_1 = C_2$, at that point $C_{total} = 0.5\ C_1$. For symmetric capacitor, the aggregate capacitance esteem levels with a large portion of the estimation of a solitary anode. For unbalanced capacitors, one of the cathodes ordinarily has a higher capacitance esteem than the other. On the off chance that C_1 is much, much greater than C_2, at that point $C_{total} \approx C_2$. Along these lines, with uneven cathodes the aggregate capacitance might be around equivalent to the littler terminal. Ultra capacitors in view of electrochemical twofold layer capacitance (EDLC) are electrical vitality stockpiling gadgets that store and discharge vitality by nano scopic charge partition at the electrochemical interface between a cathode and an electrolyte. As the vitality put away is conversely corresponding to the thickness of the twofold layer, these capacitors have a to a great degree high vitality thickness contrasted with customary dielectric capacitors. This straightforward Helmholtz EDL demonstrate was additionally changed by Gouy and Chapman with the thought of a nonstop appropriation of electrolyte particles (the two cations and anions) in the electrolyte arrangement, driven by warm movement, which is alluded to as the diffuse layer. In any case, the Gouy–Chapman show prompts an overestimation of the EDL capacitance. The capacitance of two isolated varieties of accuses expands conversely of their partition remove, consequently a substantial capacitance esteem would emerge on account of point charge particles near the anode surface. Afterward, Stern consolidated the Helmholtz display with the Gouy Chapman model to unequivocally perceive two districts of particle circulation, the inward locale called the minimal layer or the Stern layer and the diffuse layer. The structure of the electrolytic twofold layer and ingested intermediates in anode forms in fluid arrangements. Research endeavors have concentrated on expanding the vitality and power densities of Super capacitors by expanding the surface zone of permeable anodes and fitting their morphology or pore measure dissemination. Wang et al. introduced general scientific definitions for recreating electric twofold layer capacitors (EDLCs) with three-dimensional requested structures. A general arrangement of limit conditions was inferred keeping in mind the end goal to represent the Stern layer without reproducing it in the computational area. These limit conditions were legitimate for planar, barrel shaped, and round cathode particles or pores. They led the recreations of EDLCs with as intricate geometries as could be expected under the circumstances while thoroughly representing both the Stern and diffuse layers. The model additionally all the while accounted for 3D terminal morphology as well as for limited particle measure and field dependent electrolyte dielectric permittivity. It was utilized to loyally reproduce the mind-boggling structure of an EDLC anode comprising of requested bimodal mesoporous carbon, highlighting both macro pores and meso pores. Areal and gravimetric capacitances were anticipated in light of non-solvated and solvated particle distances across. These two cases set the upper and lower limits for the anticipated capacitances. The capacitances anticipated utilizing non-solvated particle width were observed to be in great concurrence with exploratory information revealed in the writing. All surfaces added to the general capacitance of the EDLCs. The gravimetric capacitance of various bimodal carbons expanded directly with expanding particular surface region comparing to steady areal capacitance.

15.5.2. Electric Double Layer Capacitors

Electric twofold layer capacitors (EDLCs) are characterized as a kind of physical battery. The EDLC has a couple of polarizable terminals with authority cathodes, a separator, and an electrolyte arrangement. Figure 15.3 demonstrates the perfect lab-scale development of an EDLC cell. The capacitor is charged and the electrical vitality put away in the capacitor is released at loads. A diagram on the advancement, history and EDLC attributes created for control utilizes "UP-Cap" is depicted in short in an audit paper by S. Nomoto et al [114]. EDLCs in view of carbon anodes has been utilized for memory go down gadget since 1978 for some, electrical machines like VCRs, camera, and so forth. In 1980s, the EDLCs were utilized for the vitality source to drive wrist watches with sun powered cells. In 1990s, they were utilized as actuator move down hotspots for toys, electric apparatuses, home gear and so on. As of late, EDLCs with higher capacitances are being worked on for higher electric power sources in electric vehicle frameworks and electric power stockpiling frameworks. Another fascinating audit on carbon materials that are utilized as anodes in EDLCs is composed by E. Frackowiak and F. Béguin; they have examined about the diverse types of carbon, for example, carbon aerogels, actuated carbon and carbon nanotubes being utilized as the anodes in EDLCs. Run of the mill electrochemical reaction of carbon nanotube in fluid electrolyte where the symmetrical and rectangular cyclic voltammetry bend is a trademark for twofold layer capacitance. Hypothetically, particular capacitance of an actuated carbon is straightforwardly relative to the particular surface zone. Be that as it may, as a general rule it doesn't occur. It was accounted for that some initiated carbons with littler surface territory give a bigger particular capacitance than those with a bigger surface zone; for example, M-30 with a BET surface region of 2571 m^2/g gave the particular limit of 62.9 Fg^{-1} while M-30 with a BET surface zone of 2130 m^2/g gave a particular limit of 100 Fg^{-1} [115].

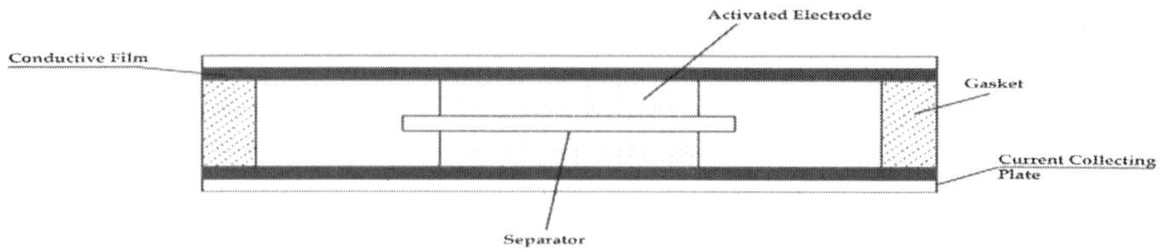

Figure 15.3: Simple construction of EDLC. (Source: Jayalakshmi, M. & Balasubramanian, K. (2008). Non-Ferrous Materials Technology Development Centre (NFTDC), Kanchanbagh Post, Hyderabad- 500058, India. *Simple Capacitors to Super capacitors – An Overview*).

Because of physical properties of enacted carbons on the conduct of EDLCs, it was accounted for by M. Nakamura et al that the rest potential was not affected by the forerunners (phenolic pitch, mineral oil, coconut shell and coal) or assembling techniques and the electrochemical qualities of the enacted carbons rely upon the oxygen content and the convergence of acidic surface practical gatherings. For a progression of nonporous enacted carbons arranged from bituminous coal, it was discovered that the gravimetric capacitance in KOH electrolyte depended not just at first glance territory of the actuated carbons yet in addition on the higher oxygen content. Notwithstanding, it was demonstrated that the electrochemical capacitance of nanoporous carbons likewise relies upon the electrolyte; carbon nanotubes with micro pores have higher capacitance in fluid electrolytes while carbon aerogel with macro pores have higher capacitance in ionic fluids. This reality infers that the particular capacitance of permeable carbon relies upon its coordinating degree to the connected electrolytes instead of its general pore volume. In another uniquely intriguing work, 25 tests of permeable carbon materials were set up from three distinct forerunners and initiated in soluble base arrangement. The specimens synthetically initiated with KOH were found to have higher oxygen content than the NaOH; a great connection between particular capacitance and CO-sort oxygen bunches has been discovered. A detailed survey on carbon properties and their part in Super capacitors was distributed by A. G. Pandolfo and A. F. Hollenkamp; they presumed that components like carbon structure, pore estimate, molecule measure, electrical conductivity and surface functionalities likewise impact the capacitance other than the surface territory. Twelve enacted carbons were all around portrayed for the variety of particular capacitance with current thickness and oxygen substance and it diminishes with expanding current thickness due to the protection in pores because of the preventing of particle move in the arbitrarily associated micro pores.

Different sorts of carbons that have been attempted as anodes in EDLCs can be condensed as: halloysite templated permeable carbon, multi-expanded permeable carbon nanofibers, nanostructured graphite, bamboo based actuated carbon, woven C-fabric, carbon nanotubes, nanostructured mesoporous carbon, carbon nanotube/felt composite terminal, cresol formaldehyde based carbon aerogel, mesoporous carbon composite (carbon nanofibers or permeable carbon), permeable carbon from thermoplastic and MgO forerunners, sodium oleate adjusted initiated carbon aerogel, TiC determined nanoporous carbon, silica MCM-48 templated mesoporous carbon, electro spun enacted carbon nanofibers, silica MCM-48 and SBA-15 templated mesoporous carbon, carbon nanofiber and PEDOT-PSS bilayer frameworks, carbon blacks, vegetable/wood based initiated carbons, actuated novoloid strands, enacted needle coke from coal tar pitch, fullerene-residue, Nomex-inferred enacted carbon filaments, actuated carbon nanotubes, poly pyrrole and graphite composite, MSP 20 enacted carbon: carbon dark: PTFE, actuated carbon-carbon nanotube composite permeable film, polymer/carbon composites, mesoporous carbon circles, pyrolyzed carbons from graphite oxides and ketjen dark/carbon nanotube, Multi-walled and single-walled carbon nanotubes, peeled carbon filaments, carbon nanotubes or enacted carbon, CNT exhibit and twofold walled CNT or actuated carbon. A far-reaching survey on cathodes in light of carbon nanotubes and enacted carbon materials for super capacitor application has been distributed as of late by V. V. N. Obreja. The capacitance esteems revealed by each creator rely upon the inception and combination of carbon also the sort of electrolyte used to absorb the terminal material [116].

15.5.2.1. Graphene based Electric Double Layer Capacitor

One graphene based super capacitor utilizes bended graphene sheets that don't stack confront to face, framing meso pores that are open to and wet table by ecologically cordial ionic electrolytes at voltages up to 4 V. They have a particular vitality thickness of 85.6 Wh/kg (308 kJ/kg) at room temperature equaling that of a regular nickel metal hydride battery, however with a power thickness 100–1000 times more noteworthy. The two-dimensional structure of graphene enhances its charging and releasing. Charge transporters in vertically arranged sheets can rapidly relocate into or out of the more profound structures of the terminal, in this manner expanding streams. Such capacitors might be reasonable for 100/120 Hz channel applications, which are inaccessible by Super capacitors utilizing different carbons.

The main use of graphene as a dynamic material for an EDLC terminal was accounted for by Vivek chand et al. in mid-2008. Keeping in mind the end goal to create vitality stockpiling gadgets with high power and vitality densities, terminals should hold all around characterized pathways for productive ionic and electronic transport. Choi et al. show elite Super capacitors by building a three-dimensional (3D) macro porous structure that comprises of artificially adjusted graphene (CMG). These 3D macro porous terminals, to be specific, emblazoned CMG (e-CMG) films, were manufactured by utilizing polystyrene colloidal particles as a conciliatory layout. Moreover, for a further capacitance help, a thin layer of MnO_2 was furthermore stored onto e-CMG. The permeable graphene structure with an expansive surface zone encourages quick ionic transport inside the anode while safeguarding not too bad electronic conductivity and subsequently invests MnO_2 or e-CMG composite cathodes with magnificent electrochemical properties, for example, a particular capacitance of 389 F/g at 1 A/g and 97.7% capacitance maintenance upon a present increment to 35 A/g. Also, when the MnO_2 or e-CMG composite terminal was unevenly gathered with an e-CMG cathode, the collected full cell indicates momentous cell execution: a vitality thickness of 44 Wh/kg, a power thickness of 25 kW/kg and a great cycle life. Yoo et al. detailed an "in-plane" creation approach for ultrathin Super capacitors in view of anodes contained perfect graphene and multilayer decreased graphene oxide. The in-plane configuration is clear to actualize and productively abuses the surface of each graphene layer for vitality stockpiling. The open engineering and the impact of graphene edges empower even the slenderest of gadgets, produced using as-grown 1–2 graphene layers, to achieve particular limits up to 80 μFcm^{-2}, while substantially higher that is about 394 μFcm^{-2} particular limits are seen in multilayer lessened graphene oxide terminals. The exhibitions of gadgets with unblemished and in addition thicker graphene-based structures are analyzed utilizing a mix of trials and model computations. The exhibited all strong state Super capacitors give a model to a wide scope of thin-film-based vitality stockpiling gadgets. Table 15.1 demonstrates an execution assessment and examination of the G and RMGO 2D "In-plane" Super capacitors. a N is the number of layers; T is the thickness of the cathode. The capacitance esteems are accounted for the best execution got utilizing the CD bends with current thickness of 281 n Acm^{-2} for RMGO and 630 $mAcm^{-2}$ for G. b Electrode capacitance changed over from the gadget capacitance expecting awry capacitor. c Normalized by the terminal mass. d Normalized by one terminal's geometrical zone. e Normalized by one anode's interface territory. f Calculated utilizing the mass, the geometrical region, and the particular region of one side of graphene (1310 m^2 g^{-1}).

Table 15.1: Performance Evaluation of G and RMGO in Super Capacitor

Device properties					Electrode properties				
Mater	Meth-od	N	T (nm)	Capaci-tance (μF)	Mass (μg)	Geometri-cal area (cm^2)	Specific capacity		
							Fg^{-1c}	μF cm^{-2d}	μF cm^{-2e}
G	CVD	1		3.333		0.1 * 0.835		80	80
RMGO	LBL	21[f]	10	35	0.283	0.085 * 2.1	247	394	19

Ruoff aggregate has spear headed another carbon material that we call artificially altered graphene (CMG). CMG materials are produced using one-particle thick sheets of carbon, functionalized as required, and here they exhibit their execution in a ultra capacitor cell. Particular capacitances of 135 and 99 F/g in fluid and natural electrolytes, individually, have been measured. Also, high electrical conductivity gives these materials reliably great execution over an extensive variety of voltage check rates. These empowering comes about show the energizing potential for superior, electrical vitality stockpiling gadgets in view of this new class of carbon material. Figure 15.4 demonstrates a SEM picture of a CMG molecule surface, TEM picture indicating individual graphene sheets reaching out from the CMG molecule surface, low and high (inset) amplification SEM pictures of CMG molecule anode surfaces, and the schematic of a test cell gathering. CMG-based ultra capacitor cells were tried with three unique electrolytes ordinarily utilized as a part of business EDLCs [117].

A super capacitor with graphene-based anodes was found to show a particular vitality thickness of 85.6 Wh/kg at room temperature and 136 Wh/kg at 80°C (all in view of the aggregate cathode weight), measured at a present thickness of 1 A/g by Liu et al. These vitality thickness esteems are similar to those of the Ni metal hydride battery, however the super capacitor can be charged or released in seconds or minutes. The way to progress was the capacity to make full usage of the most noteworthy natural surface capacitance and the particular surface region of single layer graphene by planning bended graphene sheets that won't restack up close and personal.

The bended morphology empowers the arrangement of meso pores open to and wettable by naturally kind ionic fluids fit for working at a voltage greater than 4V. Exceptionally ridged graphene sheets (HCGS) were set up by a fast, ease, and versatile approach through the warm decrease of graphite oxide at 900°C took after by quick cooling utilizing fluid nitrogen by Yan et al. The wrinkling of the graphene sheets can fundamentally keep them from agglomerating and restacking against each other up close and personal and accordingly increment the electrolyte-open surface region. The greatest particular capacitance of 349 F g^{-1} at 2mV s^{-1} is acquired for the HCGS anode in a 6 M KOH watery arrangement. Also, the terminal shows magnificent electrochemical dependability alongside an around 8.0% expansion of the underlying particular capacitance after 5000 cycle tests. These highlights make the present HCGS material a very encouraging option for the up and coming age of elite Super capacitors.

Graphene-based materials have extraordinary potential for application in Super capacitors and other environmentally friendly power vitality gadgets. There has been much enthusiasm for graphene-based electronic gadgets on the grounds that graphene gives magnificent electrical, optical, and mechanical properties. While the Super capacitors accessible today perform well, it is for the most part concurred that there is significant degree for development (e.g., enhanced execution at higher frequencies). Along these lines, it is likely that graphene will keep on playing a primary part in super capacitor innovation, for the most part through the further improvement of porosity, surface medicines to advance wettability, and diminished between molecule contact protection. The utilization of ILs as dissolvable free electrolytes in Super capacitors licenses a high cell voltage and is the most capable methodology for expanding particular vitality. A viable application in Super capacitors requires that ILs be intended to coordinate wide ESWs to high conductivity and wide obligation temperature. The outline and combination of new nanostructures and designs in view of grapheme and current ionic fluids will be an essential errand later on.

15.5.3. Pseudo Capacitors

Electrochemical twofold layer capacitor is additionally assigned as super capacitor or ultracapacitor or pseudo capacitor [118]. At the point when metal oxides/metal oxide and carbon composite/leading polymer and carbon composite are utilized as cathodes for the development of EDLCs, the charge stockpiling system incorporates both twofold layer capacitance and pseudo capacitance which result in higher capacitance yield and the EDLCs are named as Super capacitors (SCs). One noteworthy drawback of carbon based EDLC is the lower particular put away vitality. A large portion of accessible business items have a particular vitality underneath 10 Wh/kg, while the least figure for batteries is 35–40 Wh/kg, on account of lead corrosive ones, however esteem as high as 150 Wh/kg is accessible for recliner lithium particle batteries. Metal oxides display an appealing option as an anode material as a result of high particular capacitance at low protection, perhaps making it less demanding to build high vitality, high power Super capacitors. The most valuable metal oxide known to give high capacitance is RuO_2. Unadulterated RuOxHy is a blended electron-proton conductor with a high particular capacitance from 720 to 900 F/g. Broad research into ruthenium oxide has directed for military applications, where taken a toll is probably less of an issue than it is for business wanders. Scholastic establishments have concentrated on hunting down other less expensive materials rather than RuO_2. The uncommonness of this oxide is additionally fundamental factor which redirected the scientists towards other progress metal oxides. A portion of the oxides which have been examined as SC cathode material are NiO, $Ni(OH)_2$, MnO_2, Co_2O_3, IrO_2, FeO, TiO_2, SnO_2, V_2O_5 and MoO. None of these oxides are utilized as a part of business creation of EDLCs and they are still in lab-scale examine. There are gives an account of composite cathode materials which utilize blend of maybe a couple oxides alongside carbon or polymer and explored for EDLC application. Blend of ruthenium oxide with other metal oxides as thin thwarts is the underlying endeavor by many creators to sustain the examination objective. TiO_2, MoO_3, VOx, SnO_2, WO_3 and CaO are a portion of the oxides contemplated by this course. Furthermore, there are many metal oxides and blended metal oxides have been arranged and portrayed for super capacitor application by a few creators. It is a standout amongst the most critical progressing looks into in the field of vitality preservation frameworks as of late. Vanadium penta oxide V_2O_5, specifically, has been broadly inspected as an anode material for ECs that utilization natural electrolytes. Since V_2O_5 has a humble electronic conductivity, composites with metal filaments or carbonaceous materials have been set up trying to enhance terminal execution. Indistinct, nanoporous, and mesoporous types of V_2O_5 have been integrated keeping in mind the end goal to improve the particular capacitance of the oxide and the execution was assessed in KCl electrolytes; to a specific degree, these endeavors were effective. Composite terminals of V_2O_5 thin layer on conductive materials, for example, metal strands or carbonaceous materials have as of late pulled in much consideration as high rate intercalation cathodes for pseudo capacitor applications. Examination of late reports on the change metal oxides for pseudo capacitor applications demonstrates that high particular capacitance and rate ability could be acquired, particularly when a little measure of metal oxide is consistently scattered on the conductive and permeable carbonaceous materials with a high surface region, because of the expanded electrochemical use of the metal oxide and low focus polarization of the electrolyte. Vanadium oxide electrochemically kept on carbon nanotubes has been intricately examined in natural electrolytes for super capacitor application by K.B. Kim investigate gathering.

15.5.3.1. Capacitance of Pseudo Capacitor

Notwithstanding regular electrostatic collaborations in the electrical twofold layer, speedy faradic redox responses with electron exchange on the cathode/electrolyte interface may likewise add to the charge stockpiling procedure and vitality upgrade. Since the anode potential (U) shifts relatively to the charge exchanged (Q) as in capacitor, dQ = C dU, the procedure is alluded to as pseudo-capacitance. Notwithstanding, the run of the mill faradic cause of these procedures is related with a moderate dynamic of the heterogeneous response (restricted for the most part by the dispersion of the included electrochemical species) and with a direct cycle life (associated with changes of the material structure experiencing the oxidation or lessening process). Metal oxides, for example, RuO_2, MnO_2, NiO, and also electronically directing polymers, for example, polyaniline and poly pyrrole, have been widely researched in the most recent decades. Pseudo capacitance in carbon materials has been uncovered as a proficient approach to expand their capacitance by adding this impact to the EDLC. In carbon materials, pseudo capacitance can have a few causes:

i) quick redox responses including oxygen-or nitrogen-containing surface functionalities and the electrolyte.
ii) reversible electrochemical hydrogen stockpiling occurring when a negative polarization is connected to an initiated carbon cathode.
iii) redox-dynamic electrolytes at the carbon electrolyte interface, for example, iodide or iodine, vanadium or vanadyl, quinone or hydroquinone added substance to fluid electrolytes, bromide species, and so on.

Among metal oxides, RuO_2 has been the most concentrated because of its high hypothetical particular capacitance (ca. 1400–2000 F g^{-1}). Such high estimations of capacitance are identified with three particular oxidation states available inside a 1.2 V voltage window, and to a generally high metallic conductivity (105 S cm^{-1}). Nonetheless, a fantastic electrochemical action of this compound is gotten just in parotic medium. The comparing components can be depicted as a fast-reversible electron exchange together with an electro-adsorption of protons on the particles surface, as per condition, where the Ru oxidation states can change from Ru (IV) to Ru (II): [119]

$$RuO_2 + xH^+ + xe^- \leftrightarrow RuO_{2-x}\left(OH\right)_x \quad (14)$$

Practically speaking, high particular capacitance esteems, e.g., 1170 F g^{-1} and 1340 F g^{-1}, have been come to with Ru-based frameworks. Notwithstanding, this material exhibits a moderately high cost and is earth destructive. Along these lines, the scientists have turned towards less expensive materials, for example, MnO_2, NiO, Fe_3O_4, and V_2O_5. Among them, manganese oxide has a high hypothetical capacitance of 1200 F g^{-1}, while being shabby, normally plentiful and condition agreeable. High particular capacitance esteems up to 1000 F g^{-1} have been asserted for this oxide in watery Li_2SO_4 electrolyte. The capacitance is credited to reversible redox advances including the trading of protons as well as cations with the electrolyte, and in addition the changes between Mn (III)/Mn (II), Mn (IV)/ Mn (III), and Mn (VI)/Mn (IV) inside the cathode potential window of the electrolyte. The general instrument can be communicated as in

$$MnO_2\left(OC\right)_\beta + \delta C^+ + \delta e^- \leftrightarrow MnO_{\alpha-\beta}\left(OC\right)_{\beta+\delta} \quad (15)$$

where C+ signifies the protons and salt metal cations (Li+, Na+, K+) in the electrolyte, MnO (OC) and MnO – (OC) + speak to MnO_2 H_2O in high and low oxidation states, separately. For symmetric MnO_2 Super capacitors, the voltage is commonly restricted to under 1.0 V; the capacitance blurs when the negative cathode is captivated underneath 0.0 V versus Ag/AgCl because of irreversible development of dormant Mn (II) surface species. As of late, the voltage could be stretched out up to 1.2 V with an excellent security more than 10,000 cycles for a symmetric MnO_2 super capacitor utilizing Ti (IV)- containing fluid KCl electrolyte. Since MnO_2 does not show reversible advances beneath 0.0 V versus Ag/AgCl (aq), it is more reasonable to apply it as positive cathode in a half breed framework which will be depicted in the following segment. Another great case of materials giving pseudo capacitance properties are different leading polymers polyaniline-PANI, poly pyrrole (PPy), and subsidiaries of poly thiophene (PTh). PANI has been the most considered as a result of its minimal effort, generally simple readiness, high conductivity and strength. Up until now, the announced particular capacitance of unadulterated PANI adjusted anodes fluctuates from 160 to 815 F g^{-1}. Sivakkumar et al. combined PANI nanofibres by interfacial polymerization and detailed a particular capacitance of 554 F g^{1} and a poor charge or discharge cycling dependability in a symmetric PANI/PANI cell utilizing 1.0 mol L^{-1} H_2SO_4. Unadulterated PTh and PPy display particular capacitance estimations of 260 and 480 F g^{-1} in non-watery frameworks, individually. As a matter of fact, doping or DE doping of directing polymers causes volumetric changes bringing about a poor cycling soundness. Then again, these materials exhibit moderately low power execution on account of the moderate dispersion of particles inside cathode mass. Another drawback of directing polymers is their restricted security potential window. Consequently, ebb and flow examine endeavors with leading polymers for super capacitor applications are coordinated towards half and half frameworks.

15.6. PERFORMANCE

The EDLC objective is to achieve the most elevated vitality and power densities to get the littlest part volume and weight for a given application. Solidly the objectives of Montana have been settled to 5 Wh/kg for the vitality thickness and to 5 kW/kg for the power thickness.

To meet these qualities, it is important to build capacitance thickness, to decrease the ESR and to expand the cell working voltage. The endeavors are focused on the investigation of the cathode surface and openness, the grain to grain and grain to help contacts, the separator materials, and the electrolytic deterioration voltage.

In an ordinary capacitor, vitality is put away by the expulsion of charge bearers, normally electrons, from one metal plate and storing them on another. This charge partition makes a potential between the two plates, which can be bridled in an outside circuit. The aggregate vitality put away in this form is corresponding to both the quantity of charges put away and the potential between the plates. The quantity of charges put away is basically an element of size and the material properties of the plates, while the potential between the plates is constrained by the dielectric breakdown. Diverse materials sandwiched between the plates to isolate them result in various voltages to be put away. Enhancing the material prompts higher vitality densities for any given size of capacitor.

Conversely with customary capacitors, electric double layer capacitors don't have a regular dielectric. Instead of two separate plates isolated by an interceding substance, these capacitors utilize "plates" that are in reality two layers of a similar substrate, and their electrical properties, the alleged "electrical twofold layer," result in the powerful partition of charge in spite of the vanishingly thin (on the request of nanometers) physical detachment of the layers. The absence of requirement for a massive layer of dielectric allows the pressing of "plates" with considerably bigger surface zone into a given size, bringing about their exceptionally high capacitances in functional measured bundles [120].

In an electrical twofold layer, each layer by itself is very conductive, yet the material science at the interface where the layers are viably in contact implies that no critical current can stream between the layers. In any case, the twofold layer can withstand just a low voltage, which implies that electric twofold layer capacitors appraised for higher voltages must be made of coordinated arrangement associated singular electric twofold layer capacitors, much like arrangement associated cells in higher-voltage batteries.

When all is said in done, electric twofold layer capacitors enhance stockpiling thickness using a nanoporous material, commonly enacted charcoal, set up of the ordinary protecting boundary. Actuated charcoal is a powder made up of amazingly little and "harsh" particles, which in mass shape a low-thickness volume of particles with openings between them that takes after a wipe. The general surface region of even a thin layer of such a material is ordinarily more noteworthy than a conventional material like aluminum, permitting numerous more charge transporters (particles or radicals from the electrolyte) to be put away in any given volume. The drawback is that the charcoal is replacing the enhanced encasings utilized as a part of regular gadgets, so when all is said in done electric twofold layer capacitors utilize low possibilities on the request of 2 to 3 V.

Actuated charcoal isn't the "ideal" material for this application. The charge transporters are really (as a result) very substantial – particularly when encompassed by dissolvable atoms – and are regularly bigger than the openings left in the charcoal, which are too little to acknowledge them, restricting the capacity. Late research in electric twofold layer capacitors has by and large centered around enhanced materials that offer much higher usable surface territories. Test gadgets created at MIT supplant the charcoal with carbon nanotubes, which have comparative charge stockpiling capacity as charcoal (which is practically unadulterated carbon) however are mechanically organized in a significantly more general example that uncovered a substantially more prominent appropriate surface ur"u. Other groups are trying different things with custom materials made of actuated poly pyrrole, and even nanotube-impregnated papers [121].

As far as vitality thickness, existing business electric twofold layer capacitors extend around 0.5 to 30 W'h/kg, with the institutionalized cells accessible from Maxwell Technologies appraised at 6 W'h/kg and ACT underway of 30 Wh/kg. I6JI7J 5o1" however that ACT's capacitor is really a Lithium particle capacitor, referred to likewise as a "crossover capacitor." Exploratory electric twofold layer capacitors from the MIT LEES venture have shown densities of 30 W'h/kg and have all the earmarks of being adaptable to 60 W'h/kg in the short term, while EES claims their cases will offer limits around 400 W'h/kg. For correlation, a regular lead-corrosive battery is normally 30 to 40 W'h/kg and present day lithium-particle batteries are around 160 W'h/ks. Moreover. gas has a net calorific esteem (NCV) of around 12,000 W'h/kg, which in car applications works at 207o tank-to-wheel proficiency giving a successful vitality thickness of 2,400 W'h/kg.

Furthermore, electric twofold layer capacitors offer substantially higher power thickness than batteries. Power thickness consolidates the vitality thickness with the speed that the vitality can be drawn out of the gadget. Batteries, which depend on the development of charge transporters in a fluid electrolyte, have generally moderate charge and release times. Capacitors, then again, can be charged or released at a rate that is regularly restricted by current warming of the terminals. So, while existing electric twofold layer capacitors have vitality densities that are maybe 1/1Oth that of a customary battery, their energy thickness is for the most part 10 to one-hundred times as extraordinary.

15.7. DISTINCTION BETWEEN SUPER CAPACITOR AND CAPACITOR

What makes' Super capacitors unique in relation to different capacitors sorts are the terminals utilized as a part of these capacitors. Super capacitors depend on a carbon (nanotube) innovation. The carbon innovation utilized as a part of these capacitors makes a vast surface territory with an amazingly little partition separate. Capacitors comprise of 2 metal anodes isolated by a dielectric material. The dielectric isolates the terminals as well as has electrical properties that influence the execution of a capacitor.

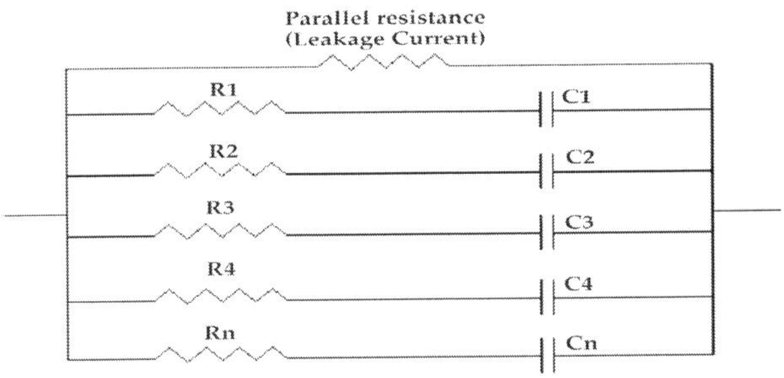

Figure 15.4: Leakage current RC circuit. (Source: *http://electrical-engineering-portal.com/super-capacitors-different-then-others-part-1*).

Super capacitors don't have a conventional dielectric material like clay, polymer movies or aluminum oxide to isolate the cathodes yet rather have a physical boundary produced using enacted carbon that when an electrical charge is connected to the material a twofold electric field is created which acts like a dielectric. The thickness of the electric twofold layer is as thin as an atom. The surface territory of the initiated carbon layer is to a great degree vast yielding a few a huge number of square meters per gram. This huge surface zone considers the assimilation of a lot of particles. The charging and discharging happens in a particle assimilation layer framed on the anodes of initiated carbon. The initiated carbon fiber terminals are impregnated with an electrolyte where positive and negative charges are framed between the anodes and the impregnant. The electric twofold layer shaped turns into an encasing until the point when a sufficiently huge voltage is connected and current starts to stream. The extent of voltage where charges start to stream is the place the electrolyte starts to separate. This is known as the decay voltage.

The twofold layers shaped on the enacted carbon surfaces can be outlined as a progression of parallel RC circuits. As appeared beneath the capacitor is comprised of a progression of RC circuits where R_1, R_2 … R_n are the interior protections and C_1, C_2, ..., C_n are the electrostatic capacitances of the actuated carbons [121].

At the point when voltage is connected current moves through each of the RC circuits. The measure of time required to charge the capacitor is dependent on the CxR estimations of each RC circuit. Clearly the bigger the CxR the more it will take to charge the capacitor. The measure of current expected to charge the capacitor is dictated by the accompanying condition

$$In = \left(\frac{V}{Rn}\right)\exp\left(-\frac{t}{Cn^{*}Rn}\right) \qquad (16)$$

In a traditional capacitor, vitality is put away by the expulsion of charge transporters, commonly electrons, from one metal plate and saving them on another. This charge partition makes a potential between the two plates, which can be outfit in an outer circuit. The aggregate vitality put away in this mold is relative to both the quantity of charges put away and the potential between the plates. The quantity of charges put away is basically an element of size and the material properties of the plates, while the potential between the plates is restricted by the dielectric breakdown. Distinctive materials sandwiched between the plates to isolate them result in various voltages to be put away. Improving the material prompts higher vitality densities for any given size of capacitor.

Interestingly with customary capacitors, electric double layer capacitors don't have a traditional dielectric. As opposed to two separate plates isolated by an interceding substance, these capacitors utilize "plates" that are in actuality two layers of a similar substrate, and their electrical properties, the alleged "electrical twofold layer," result in the viable detachment of charge regardless of the vanishingly thin (on the request of nanometers) physical partition of the layers. The absence of requirement for a massive layer of dielectric allows the pressing of "plates" with considerably bigger surface zone into a given size, bringing about their remarkably high capacitances in handy measured bundles.

In an electrical twofold layer, each layer by itselfis very conductive, yet the material science at the interface where the layers are successfully in contact implies that no critical current can stream between the layers. Notwithstanding, the twofold layer can withstand just a low voltage, which implies that electric twofold layer capacitors evaluated for higher voltages must be made of coordinated arrangement associated singular electric twofold layer capacitors, much like arrangement associated cells in higher-voltage batteries.

All in all, electric twofold layer capacitors enhance stockpiling thickness using a nanoporous material, ordinarily enacted charcoal, set up of the customary protecting hindrance. Actuated charcoal is a powder made up of to a great degree little and "unpleasant" particles, which in mass frame a low-thickness volume of particles with openings between them that takes after a wipe. The general surface zone of even a thin layer of such a material is commonly more noteworthy than a customary material like aluminum, permitting numerous more charge bearers (particles or radicals from the electrolyte) to be put away in any given volume. The drawback is that the charcoal is replacing the enhanced encasings utilized as a part of customary gadgets, so all in all electric twofold layer capacitors utilize low possibilities on the request of 2 or 3 V.

Actuated charcoal isn't the "ideal" material for this application. The charge bearers are really (basically) very substantial – particularly when encompassed by dissolvable particles – and are regularly bigger than the gaps left in the charcoal, which are too little to acknowledge them, restricting the capacity. Late research in electric twofold layer capacitors has by and large centered around enhanced materials that offer much higher usable surface zones. Test gadgets created at MIT supplant the charcoal with carbon nanotubes, which have comparative charge stockpiling ability as charcoal (which is practically unadulterated carbon) yet are mechanically masterminded in a significantly more normal example that uncovered a substantially more prominent appropriate surface ur"u. Other groups are exploring different avenues regarding custom materials made of initiated poly pyrrole, and even nanotube-impregnated papers.

Regarding vitality thickness, existing business electric twofold layer capacitors run around 0.5 to 30 W'h/kg, with the institutionalized cells accessible from Maxwell Technologies appraised at 6 W'h/kg and ACT underway of 30 Wh/kg. I6JI7J 5°1" however that ACT's capacitor is really a Lithium particle capacitor, referred to likewise as a "half and half capacitor." Test electric twofold layer capacitors from the MIT LEES venture (http://remains.mit.edu/dregs/ultra capacitors.htm) have shown densities of 30 W'h/kg and give off an impression of being versatile to 60 W'h/kg in the short term, while EEStor claims their illustrations will offer limits around 400 W'h/kg. For examination, a customary lead-corrosive battery is normally 30 to 40 W'h/kg and current lithium-particle batteries are around 160 W'h/ks. Too fuel has a net calorific esteem (NCV) of around 12,000 W'h/kg, which in vehicle applications works at 207° tank-to-wheel effectiveness giving a compelling vitality thickness of 2,400 W'h/kg.

Moreover, electric twofold layer capacitors offer substantially higher power thickness than batteries. Power thickness joins the vitality thickness with the speed that the vitality can be drawn out of the gadget. Batteries, which depend on the development of charge bearers in a fluid electrolyte, have generally moderate charge and release times. Capacitors, then again, can be charged or released at a rate that is ordinarily constrained by current warming of the cathodes. So, while existing electric twofold layer capacitors have vitality densities that are maybe 1/10th that of a regular battery, their energy thickness is by and large 10 to 100 times as incredible.

CHAPTER 16

TAXONOMY AND CHARACTERISTICS OF SUPER CAPACITORS

CONTENTS

16.1. Pseudo Capacitors 201
16.2. Hybrid Capacitors 203
16.3. High Capacitance Electrodes 205
16.4. Quantitatively Modeling 208
16.5. Measurement Of Capacitance Of Super Capacitors 210

In view of current R&D patterns, Super capacitors can be separated into three general classes: electrochemical twofold layer capacitors, pseudo capacitors, and cross breed capacitors. Its extraordinary instrument for putting away charge portrays each class. These are, separately, non-Faradaic, Faradaic, and a blend of the two. Faradaic forms, for example, oxidation-decrease responses, include the exchange of charge amongst cathode and electrolyte. A non-Faradaic instrument, by differentiate, does not utilize a substance system. Or maybe, charges are conveyed on surfaces by physical procedures that don't include the manifesting the deciding moment of compound bonds.

This area will display a diagram of every single one of these three classes of Super capacitors and their subclasses, recognized by cathode material. A graphical scientific classification of the diverse classes and subclasses of Super capacitors is introduced underneath in Figures 16.1 and 16.2

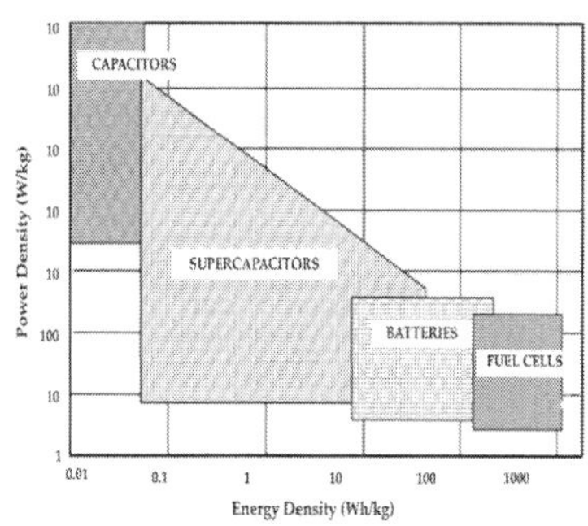

Figure 16.1: Energy vs. power plot for energy storage devices. (Source: *http://www.mpoweruk.com/images/ragone_alternatives.gif*).

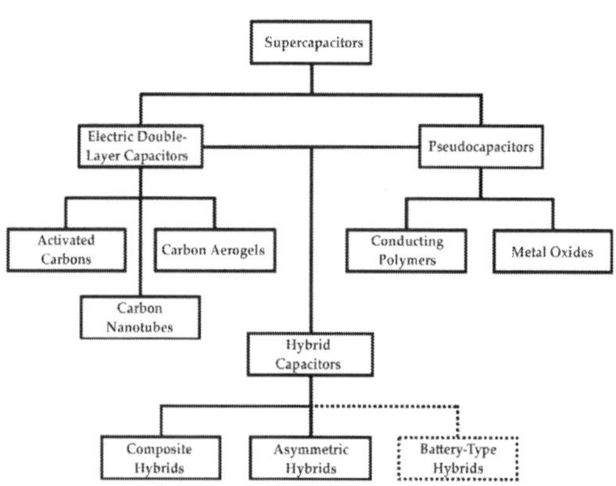

Figure 16.2: Taxonomy of super capacitors. (Source: *https://myassignmenthelp.com/mah_cms/uploads/2-1476854302.png*).

16.1. PSEUDO CAPACITORS

Rather than EDLCs, which store charge electrostatically, pseudo capacitors store charge Faradaic ally through the exchange of charge amongst anode and electrolyte. This is expert through electro sorption, lessening oxidation responses, and intercalation forms. These Faradaic procedures may enable pseudo capacitors to accomplish more prominent capacitances and vitality densities than EDLCs. There are two terminal materials that are utilized to store charge in pseudo capacitors, directing polymers and metal oxides [122].

16.1.1. Conducting Polymers

Leading polymers have a moderately high capacitance and conductivity, in addition to a generally low ESR and cost contrasted with carbon-based terminal materials. Specifically, the n/p-sort polymer arrangement, with one adversely charged (n-doped) and one decidedly charged (p-doped) directing polymer anode, has the best potential vitality and power densities; in any case, an absence of effective, n-doped leading polymer materials has kept these pseudo capacitors from achieving their potential. Also, it is trusted that the mechanical weight on directing polymers amid decrease oxidation responses restricts the soundness of these pseudo capacitors through many charge-release cycles. This diminished cycling dependability has upset the advancement of directing polymer pseudo capacitors.

16.1.2. Metal Oxides

Due to their high conductivity, metal oxides have likewise been investigated as a conceivable anode material for pseudo capacitors. The larger part of significant research concerns ruthenium oxide. This is on the grounds that other metal oxides still can't seem to acquire practically identical capacitances. The capacitance of ruthenium oxide is accomplished through the inclusion and expulsion, or intercalation, of protons into its undefined structure. In its hydrous shape, the capacitance surpasses that of carbon-based and directing polymer materials. Moreover, the ESR of hydrous ruthenium oxide is lower than that of other cathode materials. Accordingly, ruthenium oxide pseudo capacitors might have the capacity to accomplish higher vitality and power densities than comparative EDLCs and leading polymer pseudo capacitors. In any case, regardless of this potential, the achievement of ruthenium oxide has been restricted by its restrictive cost. Along these lines, a noteworthy zone of research is the improvement of manufacture techniques and composite materials to lessen the cost of ruthenium oxide, without diminishing the execution.

Surface oxides are constantly present on the surface of carbons as an outcome of the readiness forms or as a "buildup" from the carbon forerunners. Figure 16.3 condenses schematically the present condition of information on surface oxygen-containing gatherings. The primary oxygenated functionalities incorporate carboxyl-, lactone-, quinone-, pyrone-, and phenolic type gatherings, etc.

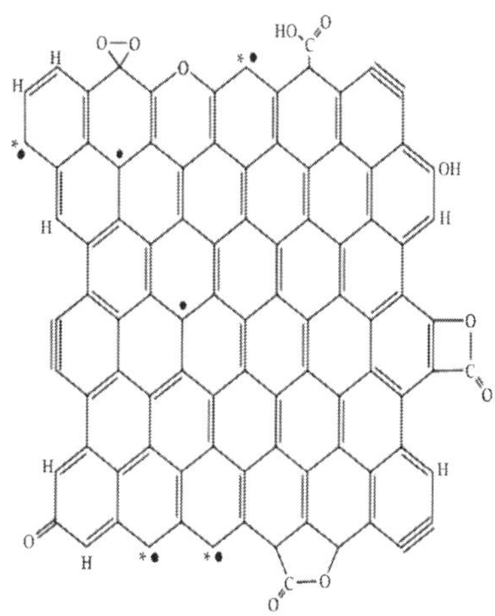

Figure 16.3: Schematic of surface oxygen containing groups. (Source: *http://pubs.rsc.org/services/images/RSCpubs.ePlatform.Service.FreeContent.ImageService.svc/ImageService/Articleimage/2016/CS/c5cs00580a/c5cs00580a-f6_hi-res.gif*).

The most reported instance of pseudo-faradic response originates from hydroquinone or quinone gatherings. For instance, a connection amongst capacitance and the measure of quinone bunches has been accounted for by Okajima et al. on oxygen plasma altered initiated carbon strands. To date, different functionalities are accounted for to display, in a less degree than quinone, some pseudocapacitive impact, for example, pyrones. In this sense, a few investigations demonstrate the connection amongst capacitance and the idea of surface functionalities. The last can be dictated by temperature-modified desorption (TPD) examination, as oxygenated functionalities on carbons are thermally deteriorated into CO and CO_2. CO develops at higher temperatures as a result of the disintegration of phenol, quinone, and ethers. Then again, the advancement of CO_2 happens at bring down temperatures relating to the decay of carboxylic gatherings or lactones. As appeared in Figure 16.3, the particular capacitance gives a decent connection the centralizations of CO-desorbed gatherings. All things considered, there is no straight relationship to the measure of oxygenated bunches desorbing as CO_2 and to the aggregate number of useful gatherings. Therefore, it is for the most part acknowledged that CO sort bunches make positive commitments to the aggregate capacitance, while CO_2 desorbing ones are proposed to impede the relocation of electrolyte into carbon pores (Figure 16.4).

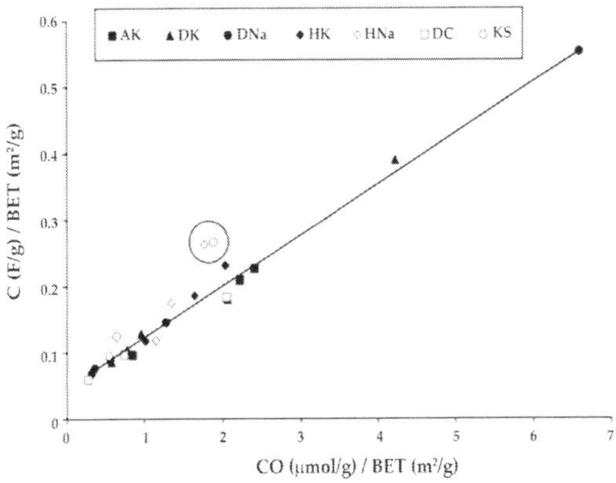

Figure 16.1: Relation between evolved C and CO. (Source: *https://www.google.com/search?client=firefox-b-ab&biw=1366&bih=695&tbm=isch&sa=1&ei=Yf1BWsXxLcWaU972q7gJ&q=relation+between+evolved+C+and+CO+of+Super capacitors&oq=relation+between+evolved+C+and+CO+of+Super capacitors&gs_l=psy-ab.3...40664.55748.0.56162.69.40.3.0.0.0.829.6836.3-10j2j3j1.16.0....0...1c.1.64.psy-ab..54.1.509...0i7i30k1.0.NCLA-WAN888#imgrc=sW_rE4d0pfJ5-M).*

16.2. HYBRID CAPACITORS

Hybrid capacitors endeavor to abuse the relative focal points and alleviate the relative weaknesses of EDLCs and pseudo capacitors to acknowledge better execution attributes. Using both Faradaic and non-Faradaic procedures to store charge, mixture capacitors have accomplished vitality and power densities more noteworthy than EDLCs without the penances in cycling strength and moderateness that have restricted the achievement of pseudo capacitors. Research has concentrated on three distinct sorts of half and half-capacitors, recognized by their terminal arrangement: composite, awry, and battery-sort separately.

16.2.1. Composite

Composite cathodes coordinate carbon-based materials with either directing polymer or metal oxide materials and join both physical and synthetic charge stockpiling instruments together in a solitary anode. The carbon-based materials encourage a capacitive twofold layer of charge and furthermore give a high-surface-territory spine that builds the contact between the stored pseudocapacitive materials and electrolyte. The pseudocapacitive materials can additionally build the capacitance of the composite terminal through Faradaic responses. Composite anodes built from carbon nanotubes and poly pyrrole, a leading polymer, have been especially fruitful. A few tests have exhibited that this terminal can accomplish higher capacitances than either an unadulterated carbon nanotube or unadulterated poly pyrrole polymer based anode. This is credited to the openness of the entrapped tangle structure, which permits a uniform covering of poly pyrrole and a three-dimensional dissemination of charge. In addition, the auxiliary uprightness of the snared tangle has been appeared to constrain the mechanical anxiety caused by the inclusion and expulsion of particles in the kept poly pyrrole. Thusly, not at all like leading polymers, these composites have possessed the capacity to accomplish a cycling solidness tantamount to that of EDLCs .[123]

16.2.2. Asymmetric Hybrids

Asymmetric Hybrids and halves consolidate Faradaic and non-Faradaic forms by coupling an EDLC anode with a pseudo capacitor terminal. Specifically, the coupling of an actuated carbon negative anode with a leading polymer positive terminal has gotten a lot of consideration. As examined before, the absence of a proficient, contrarily charged, leading polymer material has restricted the achievement of directing polymer pseudo capacitors. The usage of a contrarily charged, initiated carbon anode endeavors to go around this issue. While leading polymer terminals, for the most part, have higher capacitances and lower protections than enacted carbon anodes, they additionally have brought down greatest voltages and less cycling security. Awry crossover capacitors that couple these two terminals alleviate the degree of this tradeoff to accomplish higher vitality and power densities than practically identical EDLCs. Additionally, they have preferable cycling security over equivalent pseudo capacitors.

16.2.3. Type of Battery

Like asymmetric hybrids, battery-sort crossovers couple two distinct terminals; be that as it may, battery-sort mixtures are extraordinary in coupling a super capacitor anode with a battery cathode. This particular setup mirrors the interest for higher vitality Super capacitors and higher power batteries, consolidating the vitality qualities of batteries with the power, cycle life, and energizing circumstances of Super capacitors. Research has concentrated basically on utilizing nickel hydroxide, lead dioxide, and LTO or we can say it $Li_4Ti_5O_{12}$ as one anode and initiated carbon as the other. In spite of the fact that there is less exploratory information on battery sort crossovers than on different sorts of Super capacitors, the information that is accessible proposes that these half and halves might have the capacity to conquer any hindrance amongst Super capacitors and batteries. Regardless of the promising outcomes, the general accord is that more research will be important to decide the maximum capacity of battery-sort mixtures [124].

16.3. HIGH CAPACITANCE ELECTRODES

Early examinations on super capacitor materials basically included carbon based intensifies that show twofold layer sort capacitance. Conspicuous among these materials were actuated carbons and carbon gels, which showed very reversible surface adsorption/desorption of particles yet were constrained in vitality thickness by virtue of the surface wonder (capacitances ran in the vicinity of 50 and 130 F/g in watery electrolytes). It was normal that expanding surface territories would prompt an expansion in the gravimetric capacitance and subsequently vitality thickness. This was endeavored by making porosity in the carbons through actuation of the carbons or by utilization of formats, and in addition creating the carbons from change metal carbides or even silicon carbide by specifically scratching of the electropositive metal or semiconductor in acidic environments. Following this approach, TiC inferred nanoporous materials have been appeared to display vitality densities of 10.8 Wh/l (120 F/g capacitance) at a voltage of 2.7 V. It was however watched that the twofold layer sort capacitance did not straightforwardly scale with surface territory in many cases because of the pores being too little for electrolyte to enter. Examinations concerning this marvel prompt the conclusion that an ideal porosity of the span of two solvated particles was preferably important to get most extreme capacitance. A fascinating turn to this understanding originated from the aftereffects of the work by Gogotsi et al. who demonstrated that sub-nanometer pores considered particle desolation and in this way added to a significant increment in gravimetric capacitance emerging from a littler separation of charge separation. It was additionally discovered that the pore measure required for desolation was dissolvable ward and that this marvel prompts shape mutilation in the cyclic voltammograms. Maximum capacitance has likewise been acknowledged on account of non-fluid electrolytes when the particle estimate is practically tantamount and equivalent to the pore estimate. It has been already demonstrated that expansive pores can be gotten to rapidly while the openness of littler pores are considerably more dynamically limited. Various carbons with porosities shifting in the vicinity of 9 and 14 AO were electrochemically cycled in KOH and it was resolved that carbons with 9 AO pores had generally lesser capacitances. This impact and its suggestions are yet to be however examined on account of sub-nanometer estimate pores.

Other carbon materials being considered for super capacitor applications are carbon nanotubes (CNTs), graphene, and leading polymers. CNTs were considered to hold a particular preferred standpoint over initiated and high surface zone carbon powders in that the idea of the pores was more in the mesoporous than in the microporous run. Joined with their great conductivity and direct current to voltage reaction in lithium based electrolytes, they were believed to be perfect super capacitor materials. Capacitance esteems up to 135 F/g were accounted for with multi-walled carbon nanotubes (MWNTs) utilizing fluid electrolytes over a 0.8 V window run. It has been discovered that the carbon nanotube capacitance relies upon the width of the tubes and availability of the internal direct to particles in electrolyte like perceptions on carbon based capacitors examined before. Single walled carbon nanotubes (SWNTs) are additionally appealing because of their high electrical conductivity and conceivably bigger voltage window of stability. A current report on SWNTs in a natural electrolyte of acetonitrile has indicated capacitance like business-actuated carbons. Gravimetric capacitance has likewise been seen to change in the scope of 7 and 20 F/g for SWNTs when contrasted with 9–32 F/g for enacted carbons. Despite the fact that the innate capacitance of carbon nanotubes is little, their high electronic conductivity, mechanical quality and remarkable three-dimensional morphology have made them extremely valuable layouts for high power applications wherein pseudo capacitor materials, for example, directing polymers or metal oxides are saved on the carbon nanotubes and the composite is utilized as cathode for Super capacitors.

Among such composites, the CNT or poly pyrrole framework has been widely explored. Conducting polymers, for example, poly pyrrole shows pseudo capacitance conduct through p-or n-sort doping. Capacitances as high as 192 F/g were accounted for with poly pyrrole composites with MWNTs integrated utilizing electrochemical templating. A frequently dismissed viewpoint when utilizing exceedingly conductive materials, for example, CNTs is the contact protection between the present authority and the material itself. Shaijumon et al. have demonstrated that significant change in rate capacity is seen on fitting the interface amongst CNTs and the present authority including templating. Likewise, CNTs utilizing gold contacts have been blended utilizing alumina formats and power densities as high as 48 kW/kg have been illustrated. Another remarkable advancement is the development of graphene as a conceivable super capacitor material. Ruoff et al. spearheaded the examination into age of artificially modified graphene and has as of late shown capacitances of 99 F/g and 135 F/g in non-watery and fluid electrolytes, respectively. Recently, Chen et al. have detailed capacitances as high as 205 F/g with graphene incorporated by compound lessening of graphene oxide utilizing hydrazine. Electrochemical tests were performed in a fluid electrolyte over a 1 V window. Vitality densities as high as 28.5 Wh/kg at 10 kW/kg have been exhibited in a two terminals cell. Graphene's better properties have been ascribed than higher available surface zone up to 2600 m^2/g because of absence of agglomeration, high conductivity up to 100 S/m, and incredible synthetic dependability. Capacitor execution of graphene is additionally free of the idea of porosity due to the 2D structure managed by separating of the graphite sheets, consequently bypassing the vast majority of the difficulties required with other carbon-based materials. Carbon based Super capacitors are likewise being considered for unattached, flexible applications, for example, electronic papers and move up displays. Graphene demonstrates extraordinary guarantee in light of its great mechanical properties. Graphene/polyaniline composites were made by electro polymerization of polyaniline onto graphene paper framed by layer filtration of graphene suspension. Capacitances of around 230 F/g were accounted for in sulfuric corrosive over a 1 V window. Higher capacitances of around 530 F/g at 200 mA/g were likewise announced for poly-aniline doped graphene integrated by in situ polymerization inside graphene oxide some time recently reduction [125].

16.4. QUANTITATIVELY MODELING

The portrayals in the past area demonstrate that the scientific classification of Super capacitors incorporates vitality stockpiling frameworks that depend on an extensive variety of materials and have an extensive variety of execution qualities. To help with decreasing the time and expenses for manufacture and physical experimentation, established researchers has abused quantitative displaying to foresee the execution attributes of Super capacitors. This has decided how to create Super capacitors that perform nearer to as far as possible. Specifically, noteworthy are proportional circuit models. Research in the quantitative demonstrating of Super capacitors has concentrated on utilizing equal circuit models to catch permeable cathode conduct, and for investigating observational connections between pore estimate, surface territory, capacitance, and ESR. Likewise, such models have been utilized for deciding the hypothetical furthest reaches of Super capacitors of various structures and pieces.

16.4.1. Computer/Mathematic Model

Proportional circuit models utilize scientific or PC models of major electric circuit segments, for example, resistors and capacitors, to display complex electrochemical procedures. Basic proportional circuits have for some time been utilized to foresee the execution qualities of permeable terminals. These identical circuits essentially have been connected to endeavor to catch the conduct of the twofold layer at the interface between the terminal pores and electrolyte arrangement. All the more as of late, comparable circuits have been created to catch extra Faradaic impacts saw in pseudo capacitors. The pecking order of proportional circuits used to show permeable cathodes. This chain of importance starts with a straightforward capacitor and it also includes segments each one in turn to land at the entire equal circuit for a permeable terminal. In this last proportional circuit, which is known as a transmission line, the dispersed protections speak to the ESR characteristic for each pore as the particles from the electrolyte diffuse towards the anode.

The appropriated capacitances speak to the non-Faradaic twofold layer capacitance of each pore. It is imperative to note, also, that this comparable circuit could be changed to demonstrate a permeable pseudo capacitor anode by consolidating the Faradaic pore proportionate circuit [126].

16.4.2. Relationship Connected Empirically

There has likewise been extensive research on the observational connections between pore measure, surface region, and capacitance. As talked about in past segment, in spite of the corresponding connection between surface territory and capacitance found in principle, early proof from physical analyses proposed that surface region and capacitance were uncorrelated. Two contending numerical models have been produced to clarify this disparity amongst hypothesis and trial. The principal show recommends that, due to extraordinary electro sorption conduct found in small-scale pores, the capacitance per miniaturized scale pore surface region and capacitance per outer surface territory must be ascertained independently. The second model, which is presently generally acknowledged, proposes that electrolyte particles can't diffuse into pores underneath a size edge and in this way the surface zone of those pores can't add to the capacitance.

In considering the second model, there have been endeavors to decide the ideal pore size and size appropriation expected to augment particle availability. As a culmination result, a reverse connection between pore size and ESR should have been discovered as well.

16.4.3. Theoretical Relationship

Quantitative displaying likewise has been utilized to assess the hypothetical furthest reaches of the vitality and power densities for Super capacitors. Furthermore, by deciding the constraining elements that keep Super capacitors from achieving their hypothetical point of confinement, this exploration has produced new bits of knowledge on strategies to advance super capacitor outline. While there has been reliable enthusiasm for creating enhanced cathode materials to build vitality densities, hypothetical models propose that it is the particle fixation and breakdown voltage of the electrolyte that regularly constrain the vitality densities of Super capacitors. Besides, extra research recommends that the power densities of Super capacitors can be constrained, too, by the electrolyte. Subsequently, the exploration comes about underline that the improvement of the electrolyte is as imperative as the streamlining of the anode for accomplishing vitality and power densities nearer to the hypothetical furthest reaches of Super capacitors [127].

16.5. MEASUREMENT OF CAPACITANCE OF SUPER CAPACITORS

There are a couple of ways used to measure the capacitance of super capacitors:

1. charge method; and
2. charging and discharging method.

16.5.1. Charge Method

Measurement is performed using a charge method using the following formula

$$C = \frac{t}{R} \quad (1)$$

$$t = 0.632V_0 \quad (2)$$

where V_0 is the applied voltage.

16.5.2. Charging and Discharging Method

This method is similar to the charging method except the capacitance is calculated during the discharge cycle instead of the charging cycle.

Discharge time for constant current discharge

$$t = Cx\left(V_0 - V_1\right)/I \quad (3)$$

Discharge time for constant resistance discharge

$$t = CRln\left(\frac{V_1}{V_0}\right) \quad (4)$$

Where t is the discharge time, V_0 is the initial voltage, V_1 is the ending voltage and I is the current.

Super capacitors have such huge capacitance esteems that standard measuring hardware can't be utilized to quantify the limit of these capacitors. Capacitance is measured per the accompanying strategy [128]:

1. Charge capacitor for 30 minutes at evaluated voltage.
2. Release capacitor through a consistent current load.
3. Release rate to be 1 mA/F.
4. Measure voltage drop between V_1 to V_2.

5. Measure time for capacitor to release from V_1 to V_2.
6. Ascertain the capacitance utilizing the accompanying condition:

$$C = I^{*}\left(T_2 - T_1\right)/\left(V_1 - V_2\right) \tag{5}$$

where, $V_1 = 0.7\ V_r$, $V_2 = 0.3$ Vr and here V_r is the rated voltage of capacitor.

The significance of the need to build up the electrochemical capacitor innovation has just as of late been figured it out. This has spurned countless innovations including the advancement of new materials, for example, graphene, which indicates extraordinary guarantee inferable from its magnificent electronic properties. Also, different composite anodes have been created and exertion has been put into fitting the design of terminals to hold high vitality thickness at high rates. Essentially, change metal oxides have been explored broadly with moderately few reports on progress metal nitrides, which show up very encouraging also. At long last, the utilization of non-watery electrolytes and ionic fluids has prompted different new cell configurations with high vitality densities. The test be that as it may, lies in keeping up cyclability and steadiness at high rates keeping up these appealing attributes and subsequently inquire about endeavors are engaged toward this path. In spite of the examination propels, unmistakably advancement of super capacitor materials is to a great degree fundamental for the change of energy quality with the preferred standpoint being a speedier access to the hid away up to 106 times charge and discharge cycles.

CHAPTER 17

SUPER CAPACITORS AS ALTERNATIVE ENERGY SOURCE AND ITS APPLICATIONS

CONTENTS

17.1. Overview..214
17.2. Materials...214
17.3. Discharge Cycle 216
17.4. Advantages 217
17.5. Disadvantages...218
17.6. Applications 219
17.7. Advantages When Super Capacitor Combined With Battery..........223
17.8. Advancement In Superconductors..225

17.1. OVERVIEW

Replacing batteries with capacitors in conjunction with novel option vitality sources turned into an applied umbrella of the Green Electricity (GEL) introduced and presented by Dr. Alexander Bell. One specific effective execution of the GEL Initiative idea was a muscle-driven independent arrangement which utilizes a multi-farad electric twofold layer capacitor (hecto and kilo farad go capacitors are presently accessible) as a halfway vitality stockpiling to control an assortment of convenient electrical and electronic gadgets, for example, MP3 players, AM/FM radios, spotlights, mobile phones, and crisis kits.tl3l As the vitality thickness of electric twofold layer capacitors is conquering any hindrance with batteries, the vehicle business is sending ultra capacitors as a trade for substance batteries.

Another innovation, the super capacitor, has developed with the possibility to empower significant advances in vitality stockpiling. Super capacitors are administered by an indistinguishable major condition from ordinary capacitors, yet use higher surface territory terminals and more slender dielectrics to accomplish more prominent capacitances. This takes into account vitality densities more prominent than those of regular capacitors and power densities more prominent than those of batteries. Thus, Super capacitors may turn into an appealing force answer for an expanding number of uses. This concise review concentrates on the distinctive sorts of Super capacitors, the significant quantitative demonstrating territories, and the eventual fate of super capacitor innovative work.

17.2. MATERIALS

Initiated Carbon, Graphene, carbon nanotubes and certain conductive polymers, or carbon aerogels, are useful for Super capacitors:

For all intents and purposes all business Super capacitors made by Panasonic, Netscape, Maxwell, Nippon Chemi-Con, Axion Power, and others utilize powdered enacted carbon produced using coconut shells. A few organizations additionally fabricate higher execution gadgets, at a critical cost increment, in view of manufactured carbon forerunners that are actuated with potassium hydroxide (KOH) [129].

- Graphene has fantastic surface region per unit of gravimetric or volumetric densities, is profoundly conductive and would now be able to be created in different labs. It won't be well before extensive volumes of Graphene are created for Super capacitors.

- Carbon nanotubes have great Nano porosity properties, enabling small spaces for the polymer to sit in the tube and go about as a dielectric. MIT's Laboratory of Electromagnetic and Electronic Systems (LEES) is looking into utilizing carbon nano tuber.
- Some polymers like the polyarenes have a oxidation reduction reaction stockpiling component alongside a high surface zone.
- Super capacitors are likewise being made of carbon aerogel. This is a one of a kind material giving to a great degree high surface zone of around 400–1000 m2lg. The terminals of aerogel Super capacitors are typically made of non-woven paper produced using carbon strands and covered with natural aerogel, which at that point experiences pyrolysis. The paper is a composite material where the carbon filaments give basic trustworthiness and the aerogel gives the required vast surface. Little aerogel Super capacitors are being utilized as reinforcement power stockpiling in microelectronics, however applications for electric vehicles are normal. The voltage of an aerogel capacitor is constrained to a couple of volts. Higher voltages will prompt ionization of the carbon, which will harm the capacitor. Carbon aerogel capacitors have accomplished 325 J/g (90 Wh/kg) vitality thickness and, 20 Wg control thickness.
- Systematic pore estimate control appeared by Y-Carbon can be utilized to build the vitality thickness by more than 100% than what is monetarily accessible.
- The organization Tartu Technologies created Super capacitors from mineral based carbon. This nonactivated carbon is combined from the metal-or metalloid carbides, e.g., SiC, TiC, Al4C3, and so forth. The orchestrated nanostructured permeable carbon, frequently called Carbide Derived Carbon (CDC), has a surface zone of around 400 m^2/g to 2000 m^2/g with a particular capacitance of up to 100 F/mL (in natural electrolyte). They guarantee a super capacitor with a volume of 135 mL and 200 g weight having 1.6 kF capacitance. The vitality thickness is more than 47 kJ/L at 2.85 V and power thickness of more than 20 W/g.

Once an exploration group built up a paper battery with adjusted carbon nanotubes, intended to work as both a lithium-particle battery and a super capacitor (called bacitor), utilizing an ionic fluid, basically a fluid salt, as the electrolyte. The sheets can be rolled, bent, collapsed, or cut into various shapes with no loss of respectability or proficiency, or stacked, similar to printer paper (or a Voltaic heap), to support add up to yield. Further, they can be made in an assortment of sizes, from postage stamp to broadsheet. Their lightweight and minimal effort make them alluring for versatile hardware, air ship, autos, and toys, (for example, display flying machine), while their capacity to utilize electrolytes in blood make them possibly helpful for medicinal gadgets, for example, pacemakers. What's more, they are biodegradable.

17.3. DISCHARGE CYCLE

The charge time of a super capacitor is 1–10 seconds. The charge trademark is like an electrochemical battery and the charge current is, to an expansive degree, constrained by the charger's present dealing with capacity. The underlying charge can be made quick, and the fixing charge will take additional time. Arrangement must be made to confine the inrush current while charging a purge super capacitor as it will suck up everything it can. The super capacitor isn't liable to cheat and does not require full-charge identification; the current just quits streaming when full (Figure 17.1).

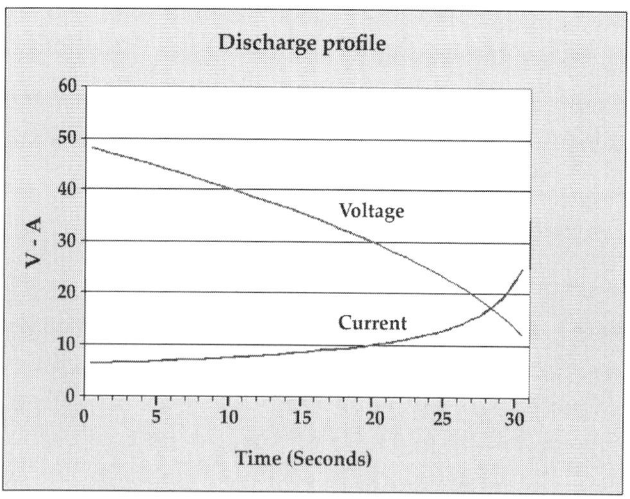

Figure 17.1: Discharge cycle of super capacitor. (Source: *http://batteryuniversity.com/index.php/learn/article/whats_the_role_of_the_super capacitor*).

The super capacitor can be charged and released an essentially boundless number of times. Not at all like the electrochemical battery, which has a characterized cycle life, there is little wear and tear by cycling a super capacitor. Age is likewise kinder to the super capacitor than a battery. Under typical conditions, a blur of a super capacitor from the first 100 percent ability to 80 percent in 10 years. Applying higher voltages than determined abbreviates the life. The super capacitor is excusing in hot and chilly temperatures, favorable position that batteries can't meet similarly well.

The self-release of a super capacitor is considerably higher than that of an electrostatic capacitor and fairly higher than an electrochemical battery; the natural electrolyte adds to this. The super capacitor releases from 100 to 50 percent in 30 to 40 days [130]. Lead and lithium-based batteries, in examination, self-release around 5 percent for each month. Because of the capacitor's high number of charge-release cycles (at least millions contrasted with 200–1000 for most monetarily accessible rechargeable batteries) there are no dispensable parts amid the entire working existence of the gadget, which makes the gadget ecologically benevolent. Batteries wear out on the request of a couple of years, and their profoundly receptive synthetic electrolytes display a genuine transfer and security danger [131]. This can be enhanced by just charging under great conditions, at a perfect rate, and, for a few sciences, as occasionally as could be expected under the circumstances. Electric twofold layer capacitors can help in such manner, going about as a charge conditioner, putting away vitality from different hotspots for stack adjusting purposes and after that utilizing any abundance vitality to charge the batteries just at fortunate circumstances [132].

17.4. ADVANTAGES

It is at present being faced off regarding what battery innovation would better serve the future car markets, where both high particular vitality and power are increasingly regularly at the same time required. The change to 42V frameworks, incorporated starter-alternator, cross breed electric vehicles and absolutely electric vehicles, will make cars more battery dependent than they right now are [133].

Different preferences of electric twofold layer capacitors contrasted and rechargeable batteries are greatly low inside protection or ESR, high productivity up to 97 or 98 Volts, high yield control, to a great degree low warming levels, and enhanced wellbeing. As indicated by ITS that is Institute of Transportation Studies, Davis, CA test comes about, the particular energy of electric two-fold layer capacitors can surpass 6 kWkg at 95 Volts of productivity [134].

To fulfill this request, one approach is to enhance existing advancements, the other is to develop new ones. Toward one side of the range, Li-particle batteries advantage from the most astounding vitality densities 150 to 200 Wh/kg, and would now be able to be designed to likewise convey high power 45 Wh/kg to 1000 W/kg. Be that as it may, they endure restricted cycle-life which normally is 500 to 1000 cycles at 100% DOD. One the opposite end, carbon-carbon twofold layer capacitors have for all intents and purposes boundless life cycles, high particular power 5000 W/kg or all the more, however humble vitality densities of 3 to 6 Wh/kg [135].

We will display a group of cross breed gadgets that utilize a nanostructured $Li_4Ti_5O_{12}$ anode and an acetonitrile $LiBF_4$ electrolyte, joined with different sorts of intercalation or two-fold layer cathodes. A large portion of them exhibit a higher cycle of life and particular power than Lithium particle, and a higher particular vitality than EDLCs. The upsides and downsides of every blend will be talked about, considering particular vitality, particular power and cycle-life [136].

Some of its other advantages are that they are little debasement more than a huge number of cycles. It has a great reversibility capability. It has very low poisonous quality of materials utilized in its working and all. It has very high cycle effectiveness of at least 95 Volts [137].

17.5. DISADVANTAGES

Super capacitors accompany a few hindrances also. One hindrance is a moderately low particular vitality. The particular vitality is a measure of aggregate sum of vitality put away in the gadget partitioned by its weight [138]. While Li-particle batteries normally utilized as a part of mobile phones have a particular vitality of 100–200 Wh/kg, Super capacitors may just store commonly 5 Wh/kg. This implies a super capacitor that has a similar limit (not capacitance) as a customary battery would weigh up to 40 fold the amount. The particular vitality isn't to be mistaken for the particular power, which is a measure of most extreme yield energy of a gadget for each weight [139].

Another burden is a straight release voltage. For instance, a battery evaluated at 2.7V, when at half charge would in any case yield a voltage near 2.7 V, while a super capacitor appraised at 2.7 V at half charge would yield precisely 50% of its most extreme charge voltage – 1.35 V. This implies the yield voltage would fall underneath the negligible working voltage of the gadget running on a super capacitor, for instance a cellphone, and the gadget would need to close down before utilizing all the charge in the capacitor. An answer for this issue is utilizing DC-DC converters. This approach presents new troubles, for example, productivity and power clamor [140].

Cost is the third significant disservice of right now accessible Super capacitors. The cost per Wh of a super capacitor is more than 20 times higher than that of Li-particle batteries. Be that as it may, cost can be lessened through new advances and large scale manufacturing of super capacitor batteries [141].

Low particular vitality, straight release voltage and high cost are the principle reasons keeping Super capacitors from supplanting batteries in many applications [142].

The measure of vitality put away per unit weight is extensively lower than that of an electrochemical battery (3–5 W.h/kg for a ultracapacitor contrasted with 30–40 W.h/kg for a battery). It is likewise just around 1/10,000th the volumetric vitality thickness of gas [143].

The voltage shifts with the vitality put away. To adequately store and recuperate vitality requires modern electronic control and exchanging hardware. It has the most elevated dielectric retention of a wide range of capacitors [144].

17.6. APPLICATIONS

Since Super capacitors conquer any hindrance amongst batteries and capacitors, they might be utilized as a part of a wide assortment of uses. One intriguing application is the capacity of vitality in KERS, or dynamic stopping mechanisms Kinetic Energy Recovery System in car industry. The fundamental issue in such frameworks is building a vitality stockpiling gadget prepared to do quickly putting away a lot of vitality. One approach is to utilize an electrical generator which will change over active vitality to electrical vitality and store it in a super capacitor. This vitality can later be reused to give energy to quickening [145].

Another illustration are low-control applications where a high limit isn't basic, however a high life cycle or speedy energizing is essential. Such applications are photographic glimmer, MP3 players, static recollections (SRAM) which require a low power steady voltage source to hold data etc. [146].

Conceivable future super capacitor applications are in phones, portable PCs, electric autos and every single other gadget that at present keep running on batteries. The most energizing preferred standpoint from a useful point of view is their quick revive rate, which would imply that connecting an electric auto to a charger for a couple of minutes would be sufficient to completely charge the battery [147].

Today little size Super capacitors as gold tops from Tokin are generally utilized as upkeep free power hotspots for IC recollections and microcomputers [148]. Among recently proposed applications for extensive size Super capacitors are stack leveling in electric and cross breed vehicles and in the footing space, the beginning of motors, applications in the media transmission and power quality and dependability necessities for uninterruptable power supply (UPS) establishments. When all is said in done Super capacitors might be adjusted to the accompanying two application spaces. The first relates to the powerful applications, where the batteries have no illustrative access. The EDLCs, on account of their powerful capacity, will permit new open doors for control gadgets. All applications where brief time control tops are required can be given by these capacitors [149]. Average cases where a major current is required amid a brief span are the quick vitality administration in half breed vehicles or the beginning of substantial diesel motors. The second one relates to the low power applications, where the batteries could be more reasonable however are at the source of upkeep issues or of lacking lifetime execution. The Super capacitors, regardless of whether they are considerably greater, convey enough preferences to substitute the batteries. In this field, the UPS and additionally security establishments are the most illustrative cases [150].

17.6.1. Basic Application

The ECDL capacitors might be utilized wherever a high-power conveyance or electrical stockpiling is required. The accompanying illustrations give a review over regular super capacitor applications [151].

17.6.2. Starter

Today the vitality that is required to wrench a little or huge motor is put away in either Pb or NiCd batteries. As a result of their high inner protection, which constrains the underlying pinnacle current, they must be larger than average. The quick battery releasing and the frosty natural temperature influence vigorously their properties. The Super capacitors have a superior power conduct and a superior natural acknowledgment [152].

17.6.2.1. Hybrid Vehicle

The mean energy of a little auto is around 30 kW and its pinnacle power ought to be around 60 kW. The Super capacitors may supply the ability to the electrical vehicle required to meet the city street movement conditions. All things considered, in a large portion of the cases there is a requirement for an extra battery to guarantee a specific measure of self-sufficiency and to achieve the range necessity. ECN Netherlands is dynamic in a power device/ super capacitor venture for an electric bike [153]. The energy unit has an energy of 1.2 kW and is mounted in a trailer pulled by the bike. The super capacitor has a 15 kWs vitality. The Super capacitors were mounted on the bike supplanting the previous battery. The possibility of the power device in addition to super capacitor idea is exhibited in this undertaking [154].

17.6.3. UPS

The uninterruptable Power Supplies (UPS) may locate some prudent interests by utilizing the ECDL capacitors, because of the concealment of an inverter and to the concealment of the support. The vitality supply amid a restricted time, at a voltage significantly higher than that of batteries, is less demanding to perform with these capacitors [155].

17.6.4. Toy and GSM Applications

Another area are toy applications, where the aggregate running time is normally not longer than 10 hours. A super capacitor intended for a long time or a few 100,000 cycles isn't advanced for such application, bring down execution is altogether adequate [156]. For short terms, the biggest markets are for gadgets with less than 12 V and just around 2004 the market for gadgets which are greater than 48 V will have developed to a similar size and will give open doors for the super capacitor advertise [157].

Amid the short 0.5 ms beat of 1 A, the battery voltage drops extensively. On the off chance that it is beneath a specific point of confinement, the telephone is not any more operable. With a super capacitor, the voltage drop is decreased fundamentally and it takes any longer until the point that the basic low voltage is come to amid the beat. Fundamentally, the operation time of the telephone is expanded [158].

There is an assortment of other exceptionally intriguing applications, which additionally underscores the monetarily fascinating parts of the Super capacitors for high-control thickness applications. Extra applications might be found in:

- elevators, cranes or bed trucks in the electric transportation space.
- hand apparatuses or electric lamps.
- radars and torpedoes in the military space.
- defibrillators and heart pacemakers in the therapeutic area.
- pulsed laser and welding in the business œ memory supplies in telephones or PCs [159].

A huge assortment of late work on Super capacitors has concentrated on the utilization of non-fluid electrolytes for super capacitor applications. This is because of the squared reliance of vitality thickness on the voltage window of the cell. Fluid electrolytes are characteristically restricted by virtue of gas development because of deterioration at high voltages [160]. To defeat this, regularly utilized lithium ion battery sort electrolytes and ionic fluids have been utilized to test different oxides and carbon-based cathode materials talked about in the past sections. Among these electrolytes, tetraethyl ammonium tetra-flouro-borate (Et_4NBF_4) in acetonitrile is most usually utilized. Vitality densities of 5 Wh/kg and high-power densities of 1200 W/kg were accounted for utilizing polyaniline anodes in natural corrosive (CF_3COOH) with a supporting electrolyte of tetra-methyl ammonium methane sulfonate. A more noticeable bearing in the super capacitor field has been toward the utilization of lopsided cells utilizing a capacitor sort anode in conjunction with a battery sort cathode [161]. Following spearheading work by Amatucci et al. utilizing the lithium titanite or acetonitrile or initiated carbon cell, different materials have been also tried and appeared to have high vitality densities that is up to 10 Wh/kg. MnO_2 has likewise been tried in a 2.2 V fluid half-breed cell with actuated carbon and appeared to have vitality densities as high as 21 Wh/kg. Zheng et al. have likewise demonstrated that vitality densities of deviated cells are reliant on the anode mass proportions and the ionic grouping of the electrolyte. Consequently, terminal coordinating of the dynamic materials content is essential for hilter kilter cells and Li et al. have proposed a hypothetical model for anode coordinating approving it utilizing a $LiMn_2O_4$ or actuated carbon framework [162].

17.7. ADVANTAGES WHEN SUPER CAPACITOR COMBINED WITH BATTERY

Capacitors store electrical charge, on the grounds that the charge is put away physically, with no synthetic or stage changes occurring, the procedure is exceptionally reversible and the release charge cycle can be rehashed again and again, for all intents and purposes unbounded [163]. Electrochemical Capacitors (ECs), differently alluded to by producers in special writing as "Super capacitors" or "Ultra capacitors," store electrical charge in an electric twofold layer at the interface between a high surface region carbon anode and a fluid electrolyte. Therefore, they are additionally legitimately alluded to as electric twofold layer capacitors (ELDCs) or just twofold layer capacitors (DLCs) [164].

A basic EC can be constructed by embeddings two conductors in a beaker containing an electrolyte, for instance, two carbon poles in salt water [165]. At first there is no quantifiable voltage between the two poles, yet when the switch is shut and current is caused to spill out of one pole to the next by a battery, charge partition is normally made at each fluid strong interface. This effectively makes two capacitors that are arrangement associated by the electrolyte [166]. Voltage endures after the switch is opened vitality has been put away. In this state, solvated particles in the electrolyte are pulled in to the strong surface by an equivalent but opposite charge in the strong [167].

These two parallel areas of charge frame the wellspring of the expression "double layer." Charge partition is measured in sub-atomic measurements (i.e., barely any angstroms), and the surface territory is measured in a huge number of square meters per gram of cathode material, making five thousand Farad estimate capacitors that can be hand held [168].

Super capacitors are in this way not suited to vitality stockpiling (like a synthetic battery) over drawn out stretches of time, but instead they are great at here and now stockpiling and conveyance at around 98% of the charge placed in them. The graph represents the capability of the Super capacitor to convey high wrenching force [169].

The working voltage of an electrochemical capacitor is constrained by the breakdown capability of the electrolyte regularly 1 to 3V for every cell and is by and large much lower than that of ordinary electrostatic and electrolytic capacitors. In numerous down to earth applications, subsequently, electrochemical capacitor cells must be arrangement associated, like batteries, to meet operating voltage necessities [170].

The major contrast amongst batteries and electrochemical capacitors is that the previous store vitality in the main part of substance reactants fit for producing charge, while the last store vitality straightforwardly as surface charge [171]. Along these lines, the battery has significantly more vitality in a given volume than a capacitor. Battery release rate and in this manner control execution is constrained by the response energy and also the mass transport, while such impediments don't have any significant bearing to electrochemical capacitors developed with two enacted carbon terminals, in this way giving capacitors outstandingly high power ability amid both release and charge [172].

Most batteries show a moderately consistent working voltage in light of the thermodynamics of the battery reactants; as an outcome, it is frequently hard to gauge their State of Charge (SOC) accurately [173]. What's more, for a capacitor, its working voltage changes directly with time amid consistent current operation so that the SOC can be precisely known. Besides, the profoundly reversible electrostatic charge stockpiling instrument in ECs does not prompt any volume change like saw in batteries with electrochemical changes of dynamic masses. This volume change constrains the cyclability of batteries by and large to less than 2500 cycles while ECs have shown from several thousands to a huge number of full charge/release cycles [174].

On the off chance that we look at the state of the workmanship business beginning lighting start (SLI) battery innovation with electrochemical capacitor innovation in several important ways. Cost figures depended on the present product prices for capacitors and batteries in huge amounts, each standardized to mass utilizing dc vitality esteems and power levels at which charge and release effectiveness is no less than 90% [175].

Unmistakably the charge and release times and the cycle life [176] of capacitor and SLI lead corrosive battery innovations are very different at present, and it is normal that this circumstance isn't probably going to change within a reasonable time-frame. Indeed, even the most developed lithium particle battery innovation accessible today offers charge/release time is almost 3–5 minutes, particular power levels of 1 kW/kg, and operates at about 5000 cycles.

Subsequently the reasonableness of every innovation for specific applications is currently and will remain notably extraordinary paying little heed to likely advances that might be made in both of the advances. Along these lines, capacitors are relied upon to remain the innovation of decision in applications requiring high-control execution, as in motor wrenching [177]. It is simply impractical to conquer the central confinements of battery innovation to have them viably contend with the execution of capacitors for motor turning.

Super capacitors are a feasible expansion or alternate to the beginning heartbeat control requirements for SLI batteries as used today in motor vehicles, and especially suited to the current quickly developing Micro Hybrids for the Idle Stop Start (ISS) functionality [178].

17.8. ADVANCEMENT IN SUPERCONDUCTORS

In the progressing exertion for vitality manageability, electrochemical Super capacitors are developing as gadgets of prime significance inferable from their prevalent attributes unmatched by some other charge stockpiling gadget. These include: high power densities at generally high vitality densities and long cycle life. These attributes have just been misused in different applications, for example, control hardware, expansive scale transport frameworks containing metro prepares and transports, vitality stockpiling at irregular generators including windmills, and brilliant lattice applications. Brilliant Grid needs will incorporate network solidness, recurrence smoothing, top moving, and recurrence control, as talked about in the before articles in this issue. The future market for batteries and super capacitor stockpiling frameworks is relied upon to develop to $8 billion out of 2017. The execution of state of the workmanship in many different business Super capacitors is reasonable for various different applications.

Since Super capacitors are described by quick release rates and low support, the interest for them is anticipated to be most broad in recurrence direction applications. The other territory for significant development is in regenerative braking for network associated light rail frameworks. Expectedly, Super capacitors have been arranged into two sorts: electrochemical twofold layer capacitors (EDLCs) and pseudo capacitors in light of the Faradaic electrochemical redox instrument of charge stockpiling. The electrochemical interface between a terminal and electrolyte has for quite some time been known to carry on like a capacitor and hence the name "twofold layer" is frequently generally utilized while alluding to the capacitor marvel. The quick reaction of the interface to changes in terminal potential and high reversibility have been misused in EDLCs. Pseudo capacitors then again are gadgets in light of terminal charge stockpiling joined by exceedingly reversible Faradaic [179] electrochemical redox surface procedures. Materials focused for EDLCs have principally been high surface zone carbons while pseudo capacitor investigate has to a great extent been commanded by oxides, for example, those of ruthenium and manganese. A schematic outline for an EDLC is appeared in Figure 17.2.

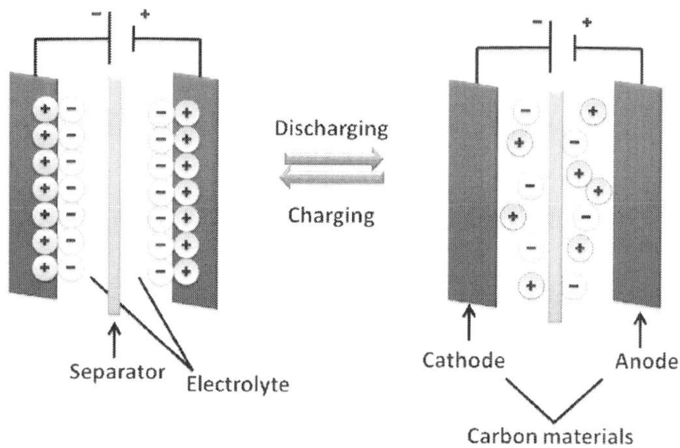

Figure 17.2: Schematic diagram of EDLC. (Source: *http://www.enerize.com/images/EDLCDiagramColor2.JPG*).

The main role behind the blossoming exertion in super capacitor gadgets is the need to conquer any hindrance between ordinary capacitors and batteries with respect to vitality/control execution. For Super capacitors to be monetarily feasible for substantial scale applications there is a requirement for more noteworthy vitality densities and lessening in cost [180]. More prominent vitality densities can be acquired by expanding either the anode capacitance or the electrolyte voltage window. The previous is accomplished by utilizing high capacitance cathode materials and the last through utilization of non-aqueous electrolytes with bigger window of electrochemical strength. This has impelled research into countless anode materials including nanocomposites to enhance capacitance to a great extent misusing the speedier electronic and ionic transport attributes of nanoscale frameworks. Expanding the voltage window notwithstanding, has been recommended as a more reasonable way to deal with enhancing the vitality thickness by virtue of the squared reliance of vitality thickness on voltage and in addition the absence of issues related with an expansion in the RC-time consistent on expanding capacitance to accomplish higher vitality thickness prompting a drop in rate ability.

REFERENCES

1. Eddahech, A., Briat, O., & Vinassa, J. M., (2013). *Energy,* 61432.
2. Saito, Y., Kanari, K., & Takano, K., (1997). *J. Power Sources,* 68451.
3. Lu, W., Yang, H., & Prakash, J., (2006). *Electrochimica. Acta,* 511322.
4. Choi, Y. S., & Kang, D. M., (2014). *J. Power Sources,* 270273.
5. Walker, W., & Ardebili, H., (2014). *J. Power Sources,* 269486.
6. Srinivasan, R., Baisden, A. C., Carkhuff, B. G., & Butler, M. H., (2014). *J. Power Sources*, 26293.
7. Kumaresan, K., Sikha, G., & White, R. E., (2008) *J. Electrochem. Soc.*, 155 A164.
8. Doyle, M., Fuller, T. F., & Newman, J., (1993). *J. Electrochem. Soc.,* 1401526.
9. Giuliano, M. R., Advani, S. G., & Prasad, A. K., (2011). *J. Power Sources, 196,* 6517.
10. Tong, W., Somasundaram, K., Birgersson, E., Mujumdar, A, S., & Yap, C., (2015). *Inter. J. Ther. Sci.,* 94259.
11. Wang, Q., Ping P., Zhao, X., Chu G., Sun, J., & Chen, C., (2012). *J. Power Sources* 208210.
12. Feng, X., Fang, M., He, X., Ouyang, M., Lu, L., Wang, H., & Zhang, M., (2014). *J. Power Sources*, 255294.
13. Richard, M. N., & Dahn, J. R., (1999). *J. Power Sources*. 79135.
14. Hatchard, T. D., MacNeil, D. D., Basu, A., & Dahn, J. R., (2001). *J.*

Electrochem. Soc., 148A755.

15. Hannoyer, H., Prince, A. A. M., Jean, M., Liu, R. S., & Wang, G. X., Mossbauer study on
16. $LiFePO_4$, (2004). Cathode material for lithium ion batteries. Hyperfine Interact., 167,767, 2006.
17. Doeff, M. M., Hu, Y. Q., Mclarnon, F., & Kostecki. R., (2003). Effect of Surface Carbon Structure on the Electrochemical Performance of LiFePO4. *Electrochem. Solid-State Lett., 6*, A207.
18. Dominko, R., Bele, M., Gaberscek, M., Remskar, M., Hanzel, D., Pejovnik, S., & Jamnik, J., (2005). Impact of the Carbon Coating Thickness on the Electrochemical Performance of LiFePO4/C Composites. *J. Electrochem. Soc., 152*(3), A607.
19. Zaghib, K., Mauger, A., Kopec, M., Gendron, F., & Julien, C. M., (2009). Intrinsic Properties of 40 nm-Sized $LiFePO_4$ Particles. *ECS Transactions, 16*(42), 31.
20. Jones, J. L., Hung, J. T., & Meng, Y. S., (2009). Intermittent X-ray Diffraction study of Kinetics of Delithiation in Nano-scale LiFePO4. *J. Power Sources, 189*(1), 702.
21. Konarova, M. & Taniguchi, I., (2010). Synthesis of Carbon-coated $LiFePO_4$ Nanoparticles with high rate Performance in Lithium Secondary Batteries. *J. Power Sources, 195*(11), 3661.
22. Malik, R., Burch, D., Bazant, M., & Ceder, G., (2010). Particle size Dependence of the Ionic Diffusivity. *Nano Lett., 10*, 4123.
23. Ellis, B., Herle, P. S., Rho, Y. H., Nazar, L. F., Dunlap, R., Perry, L. K., & Ryan, D. H., (2007). Nanostructured Materials for Lithium-ion Batteries, Surface Conductivity vs. Bulk Ion/Electron Transport. *Faraday Discuss., 134*, 119.
24. Tang, M., Huang, H. Y., Meethong, N., Kao, Y. H., Carter, W. C., & Chiang, Y. M., (2009). Model for the Particle Size, Overpotential, and Strain Dependence of Phase Transition Pathways in Storage Electrodes, Application to Nanoscale Olivines. *Chem. Mater., 21*(8), 1557.
25. Burch, D., & Bazant, M. Z., (2009). Size-Dependent Spinodal and Miscibility Gaps for Intercalation in Nanoparticles. *Nano Lett., 9*(11), 3795.
26. Mott, N. F., (1968). Conduction in Glasses Containing Transition Metal Ions. *J. Non-Cryst. Solids, 1*, 1.
27. Xu, Y., Shankland, T. J., & Duba, A. G., (2000). Phys. *Earth Planetary*

Interiors, 118, 149.

28. Ingalls, R. (1964). Electric-field Gradient Tensor in Ferrous Compounds. *Phys. Rev., 133*(3A), A787.
29. Danot, M., Afanasov, M. I., Fabritchnyi, P. B., Golub, A. S., Lenenko, N. D., Novikov, Y. N., Pokholok, K. V., & Presniakov, I. A., (2000). M¨Ossbauer Evidence for Fast Electron Hopping between Fe(II) & Fe(III) in Iron Hydroxide/Molybdenum Disulfide Lamellar Nanocomposite. *Solid State Ionics, 128*, 211.
30. Wagemaker, M., Mulder, F. M., & A. Van der Ven. (2009). The Role of Surface and Interface Energy on phase Stability of Nanosized Insertion Compounds. *Adv. Mater., 21*, 2703.
31. Zaghib, K., Mauger, A., Gendron, F., & Julien, C. M., (2008). Surface Effects on the Physical and Electrochemical Properties of Thin LiFePO4 Particles. *Chem. Mater., 20*, 462.
32. Lee, K. T., Kan, W. H., & Nazar, L. F., (2009). Proof of Intercrystallite Ionic Transport in $LiMPO_4$ Electrodes (M = Fe, Mn). J. Am. *Chem. Soc., 131*(17), 6044.
33. Van der Ven, K. Garikipati, S. Kim, & M. Wagemaker. (2009). The Role of Coherency Strains on Phase Stability in LixFePO4: Needle Crystallites Minimize Coherency Strain and Overpotential. *J. Electrochem. Soc., 156*(11), A949.
34. Reynier, Y., Graetz, J., Swan-Wood T., Rez P., Yazami R., & Fultz, B., (2004). Entropy of Li Intercalation in LixCoO2. *Phys. Rev. B, 70*(17), 174304.
35. Reynier, Y., Yazami, R., Fultz, B., & Barsukov, I., (2007). Evolution of Lithiation Thermodynamics with the Graphitization of Carbons. J. Power Sources, *165*(2), 552. [Joint Meeting of the International-Battery-Association/Hawaii Battery Conference (IBA-HBC 2006), Hilo, HI, Jan, 2006].
36. Matsui, H., Nakamura, T., Kobayashi, Y., Tabuchi, M., & Yamada, Y., (2010). Open-circuit Voltage Study on $LiFePO_4$ Olivine Cathode. *J. Power Sources, 195*, 6879.
37. Goodman, R. H., & Richardson, J. E., (1966). Use of "on Line" Computer for Mossbauer Experiments. *Rev. Sci. Instrum., 37*(3), 283.
38. Fultz, B., & Morris, J. W., (1978). Multichannel Scaling with an Eight bit Microcomputer. *Rev. Sci. Instrum., 49*(8), 1216.

39. Benedetti, M., & Fernandez, D. R., (1981). Fully Programmable Microprocessor-Controlled M¨ossbauer Spectrometer. *Rev. Sci. Instrum., 52*(9), 1406.
40. S´anchez Asseff, A. J., Boh´orquez, A., P´erez Alc´azar, G. A., & Pfannes, H. D., (1997). An Autonomous Programmable System for M¨ossbauer Spectroscopy, Accessed by a PC Program. *Hyperfine Interact., 110*, 135.
41. Malik, R., Zhou, F., & Ceder, G., (2011). Kinetics of Non-equilibrium Lithium Incorporation in LiFePO4. *Nat. Mater., 10*, 587.
42. Yamada, Chung, S. C., & Hinokuma, K., (2001). Optimized $LiFePO_4$ for Lithium Battery Cathodes. *J. Electrochem. Soc., 148*(3), A224,.
43. Prosini, P. P., Lisi, M., Zane, D., & Pasquali, M., (2002). Determination of the Chemical Diffusion Coefficient of Lithium in $LiFePO_4$. *Solid State Ionics, 148*, 45.
44. Takahashi, M., Tobishima, S., Takei, K., & Sakurai, Y., (2001). Characterization of LiFePO+4 as the Cathode Material for Rechargeable Lithium Batteries. *J. Power Sources, 97*, 508.
45. Kim, H. S., Cho, B. W., & Cho, W. I., (2004). Cycling Performance of $LiFePO_4$ Cathode Material for Lithium Secondary Batteries. *J. Power Sources, 132*, 235.
46. Mi, C. H., Zhao, X. B., Cao, G. S., & Tu, J. P., (2005). In situ Synthesis & Properties of Carbon-Coated $LiFePO_4$ as Li-ion Battery Cathodes. *J. Electrochem. Soc., 152*, A483.
47. Arnold, G., Garche, J., Hemmer, R., Strobele, S., Vogler, C., & Wohlfahrt-Mehrens, M., (2003). Fine Particle Lithium Iron Phosphate $LiFePO_4$ Synthesized by a New Low-cost Aqueous Precipitation Technique. J. *Power Sources, 119*, 247.
48. Delacourt, C., Poizot, P., Levasseur, S., & Masquelier, C., (2006). Size Effects on Carbon-free $LiFePO_4$ Powders. *Electrochem. and Solid-State Lett., 9*(7), A352.
49. Chung, H. T., Jang, S. K., Ryu, H. W., & Shim, K. B., (2004). Effects of Nano-carbon Webs on the Electrochemical Properties in $LiFePO_4/C$ Composite. *Solid State Commun., 131*, 549.
50. Singhal, S., Skandan, G., Amatucci, G., Badway, F., Ye, N., Manthiram, A., Ye, H., & Xu, J. J., (2004). Nanostructured Electrodes for Next Generation Rechargeable Electrochemical Devices. *J. Power Sources,*

129, 38.

51. Spong, D., Vitins, G., & Owen, J. R., (2005). A Solution-Precursor Synthesis of Carbon-coated $LiFePO_4$ for Li-ion Cells. *J. Electrochem. Soc., 152*, A2376.
52. Yang, M. R., Ke, W. H., & Wu, S. H., (2005). Preparation of LiFePO4 Powders by Co-precipitation. *J. Power Sources, 146,* 539.
53. Klug, H. P., & Alexander, L. E., (1974). X-ray Diffraction Procedures for Polycrystalline and Amorphous Materials. *Wiley*.
54. Huang, H., Yin, S. C., & Nazar, L. F., (2001). Approaching Theoretical Capacity of $LiFePO_4$ at Room Temperature at High Rates. Electrochem. *Solid-State Lett., 4*(10), A170.
55. Thomas-Alyea, K. E., & Newman, J., (2003). Thermal Modeling of Porous Insertion Electrodes. *J. Electrochem. Soc., 150*, A176–A192.
56. Gu, W. B., & Wang, C. Y., (2000). Thermal-Electrochemical Modeling of Battery Systems. *J. Electrochem. Soc., 147*, 2910–2922.
57. Wang, C. Y., Gu, W. B., & Liaw, B. Y., (1998). Micro-Macroscopic Coupled Modeling of Batteries and Fuel Cells. I. Model Development. *J. Electrochem. Soc., 145*, 3407–3417.
58. Verbrugge, M. K., (1995). Three-dimensional Temperature and Current Distribution in a Battery Module. *AIChE J.,* 41, 1550–1562.
59. Huang, H., & Nguyen, T. V., (1997). A Transient Nonisothermal Model for Valve-regulated Lead-Acid Batteries Under float. *J. Electrochem. Soc., 144*, 2420–2426.
60. de Vidts, P., Delgado, J., Wu, B., See, D., Kosanovich, K., & White, R. E., (1998). A Nonisothermal Nickel-Hydrogen Cell Model. *J. Electrochem. Soc., 145*, 3874–3883.
61. de Groot, S. R., & Mazur, P., (1984). Non-Equilibrium Thermodynamics. *Dover*.
62. Landau, L. D., & Lifshitz, E. M., (1959). Fluid Mechanics. *Pergamon,*.
63. Landau L. D., & Lifshitz, E. M., (1984). Electrodynamics of Continous Media. *Pergamon, Oxford.*
64. Henjes, K., & Liu, M., (1993). Hydrodynamics of Polarizable Liquids. *Ann. Physics, 223*, 243–276.
65. Liu, M., (1993). Hydrodynamic Theory of Electromagnetic fields in Continous Media. Phys. *Rev. Lett., 70*, 3580–3583.
66. Jiang, Y., & Liu, M., (1996). Dynamics of dispersive and nonlinear

media. *Phys. Rev. Lett., 77*, 1043–1046.

67. Jiang, Y. & Liu, M., (1998). Electromagnetic Force in Dispersive and Transparent Media. *Phys. Rev. E, 58*, 6685–6694.
68. Liu, M., (1995). Fluid Dynamics of Colloidal and Electric Liquids. *Phys. Rev. Lett., 74*, 4535–4538.
69. Latz., & Zausch, J., (2010). Modeling of Species and Charge Transport in Li-ion Batteries Based on Non-equilibrium Thermodynamics. *Berichte des Fraunhofer ITWM, 190*, 1–8.
70. Bazant, M. Z., Thornton, K., & Ajdari, A., (2004). Diffuse-charge Dynamics in Electrochemical Systems. *Phys. Rev. E, 70*, 021506.
71. Bazant, M. Z., & Chu, K. T., (2006). Nonlinear Electric Relaxation Around Conductors. *Phys. Rev. E, 74,* 011501.
72. Tyrrell, H. J. V., (1953). Thermal Diffusion Phenomena in Electrolytes and the Constants Involved, *National Bureau of Standards (U.S.) Circular. National Bureau of Standards, 524*.
73. ean K., Platten., (2006).The Soret Effect: A Review of Recent Experimental Results. *Journal of Applied Mechanics, 73*(1), 5–15.
74. Colclasure, M., & Kee, R. J., (2010). Thermodynamically Consistent Modeling of Elementary Electrochemistry in Li-ion Batteries. *Electrochimica Acta*.
75. Kuzminskii, Y. V., & Gorodyskii, A. V., (1988). Thermal Analysis of Electrochemical Reactions Part I. Kinetic Method of Determining Peltier heats. *J. Electroanal. Chem, 252*, 21–38.
76. Wood, C., Chmielewski, A., & Zoltan, D., (1988). Measurement of Seebeck Coefficient Using a Large Thermal Gradient. *Rev. Sci. Instrum., 59*, 951–954.
77. Wang, C. W., & Sastry, A. M., (2007). Mesoscale Modeling of Li-ion Polymer Cell. *J. Electrochem. Soc., 154*, A1035–A1047.
78. Zausch, J., Latz, A., Schmidt, S., Less, G. B., Seo, J. H., Han, S., & Sastry, A. M., (2010). Micro-scale Modeling of Li-Ion Batteries; Parameterization and Validation. to be Published.
79. Christensen, J. & Newman, J., (2004). A Mathematical Model for Lithium-Ion Negative Electrode Solid Electrolyte Interphase., *J. Electrochem. Soc., 151*, A1977–A1988.
80. Sieniutycz, S., (2003). A Synthesis of Thermodynamic Models Unifying Traditional and work-Driven Operations with Heat and Mass Exchange. *Open Sys. & Information Dynamics, 10*(1), 31–49.

81. Sieniutycz, S., (2008). An Analysis of Power and Entropy Generation in a Chemical Engine. Intern. *J. of Heat and Mass Transfer, 51*(25–26), 5859–5871.

82. Sieniutycz, S., (2009a). Dynamic Programming and Lagrange Multipliers for Active Relaxation of Resources in Non-equilibrium Systems.

83. Sieniutycz, S., (2009b). Applied Mathematical Modeling, *33*(3), 1457–1478.

84. Complex Chemical Systems with Power Production Driven by Mass Transfer. Intern. *J. of Heat and Mass Transfer, 52* (10), 2453–2465.

85. Sieniutycz, S., & Jeowski, J., (2009). Energy Optimization in Process Systems. Oxford. Elsevier.

86. Sieniutycz, S., & Kuran, P., (2005). Nonlinear Models for Mechanical Energy Production in Imperfect Generators Driven by Thermal or Solar Energy.

87. Sieniutycz, S., & Kuran, P., (2006). Modeling Thermal Behavior and Work flux in finite-rate Systems with Radiation. Intern. *J. Heat Mass Transfer, 48*(3–4), 719–730.

88. Tsirlin, A. M., Kazakov, V., Mironova, V. A., & Amelkin, S. A., (1998). Intern. *J. Heat and Mass Transfer, 49*(17–18), 3264–3283.

89. Finite-time Thermodynamics: Conditions of Minimal Dissipation for Thermodynamic Process. *Physical Review E, 58*(1), 215–223.

90. Wierzbicki, M., (2009). Optimization of SOFC Based Energy System Using Aspen Plus^{TM}. Thesis (MsD) Supervised by S. Sieniutycz (Faculty of Chemical and Process Engineering, Warsaw University of Technology) and J. Jewulski (Laboratory of Fuel Cells, Institute of Energetics, Warsaw.

91. Hirschen Hofer, J. H., Stauffer, D. B., & Engleman, R. R., (1994). Fuel Cells A Handbook (Revised 3), U.S. Department of Energy.

92. Kordesch, K. & Simader, G. Fuel Cells and Their Applications, VCH, New York,1996. J. Apple by, Fuel cell technology: status and future prospects. Energy,21 (1996) 521–653.

93. Vielstich, W., Gasteiger, H., Lamm, A., (Ed.). (2003). Handbook of Fuel Cells–Fundamentals, Technology, Applications, Wiley.

94. Appleby, J. (1993). Characteristics of Fuel Cell Systems. In Fuel Cell Systems (Eds.) Blomen, L. J. M. J. & Mugerwa, M. N.), Plenum Press, New York.

95. K. Wark Jr., (1995). Advanced Thermodynamics for Engineers, McGraw-Hill, New York.
96. Haynes, C., (2001). Clarifying Reversible Efficiency Misconceptions of High Temperature Fuel Cells in Relation to Reversible Heat Engines. *J. Power Sources, 92,* 199–203.
97. Lutz, E. A., et al., (2002). Thermodynamics comparison of fuel cells to Carnot cycle. *Int. J. Hydrogen Energy, 27*, 1103–1111.
98. Wright, S. E., (2004). Comparison of the Theoretical Performance Potential of Fuel Cells and Heat–Engines, *J. Renewable Energy, 29*, 179–195.
99. Hassanzadeh, H., & Mansouri, S. H., (2005). Efficiency of Ideal Fuel Cell and Carnot Cycle from a Fundamental Perspective, Proceedings of IMechE, Part A: *J. Power Energy, 219.*
100. Xianguo Li., (January 1997). Fuel cells and thermodynamic performance, 8th Annual international Conference ASME, *Energy Week,* 28–30.
101. Xianguo Li, (2002). Fuel Cells–the Environmentally Friendly Energy Converter and Power Generator, Int. *J. Global Energy Issues, 17*,Nos. 1/2.
102. Barendrech, E., (1989). Electrochemistry of Fuel Cells. *In Fuel Cell Systems* (Eds. L. J. M. J. Blomen & M. N. Mugerwa), Plenum Press, New York.
103. Baschuk, J. J., & Xianguo Li., (2001). Carbon Monoxide Poisoning of Proton Exchange Membrane Fuel Cells, Int. *J. Energy Res., 25,* 695–713.
104. Schrodinger, E., (1989). Statistical Thermodynamics, Dover, New York. This is a short Monograph written by the one of the Pioneers of Quantum Physics, the co-winner of the 1933 Nobel Prize in Physics. It is based on a set of lectures delivered to the Dublin Institute for Advanced Studies in 1944, and was first published in 1946.
105. Shapiro, H., (1953). The Dynamics and Thermodynamics of Compressible Fluid Flow, Vols. I and II, John Wiley, New York. This classic two volume set has a comprehensive treatment of the subject of its title. It has numerous worked example problems, and is written from a careful engineer's perspective.
106. Smith, J. M., van Ness, H. C., & Abbott, M., (2004). Introduction to Chemical Engineering Thermodynamics, Seventh Edition, McGraw-

Hill, New York. This is probably the most common undergraduate text in thermodynamics in chemical engineering. It first appeared in 1959. It is rigorous and has went through many revisions.

107. Sommerfeld, (1956). Thermodynamics and Statistical Mechanics, Lectures on Theoretical Physics, Vol. V, Academic Press, New York. This is a compilation of the author's lecture notes on this subject. The book reflects the author's stature of a leader of theoretical physics of the twentieth century who trained a generation of students (e.g., Nobel laureates Heisenberg, Pauli, Debye, & Bethe). The book gives a fine description of classical thermodynamics with a seamless transition into quantum and statistical mechanics.
108. Tester, J. W., & Modell, M., (1997). Thermodynamics and Its Applications, Third Edition, Prentice Hall, Upper Saddle River, New Jersey. First appearing in 1974, this entry level graduate text in thermodynamics is written from a chemical engineer's perspective. It has a strong mathematical development for both classical and statistical thermodynamics.
109. Truesdell, C. A., (1980). The Tragicomic History of Thermodynamics, 1822–1854, Springer Verlag, New York. This idiosyncratic monograph has a lucid description of the history of nineteenth century thermal science. It is written in an erudite fashion, and the reader who is willing to dive into a difficult subject will be rewarded for diligence by gain of many new insights.
110. Truesdell, C. A., (1984), Rational Thermodynamics, Second Edition, Springer-Verlag, New York. This is an update on the evolution of classical thermodynamics in the twentieth century. The book itself first appeared in 1969. The second edition includes additional contributions by some contemporaneous leaders of the field.
111. Turns, S. R., (2000). An Introduction to Combustion, Second Edition, McGraw-Hill, Boston. This is a senior-level undergraduate text on combustion which uses many notions from thermodynamics of mixtures. It first appeared in 1996.
112. Turns, S. R., (2006). Thermodynamics: Concepts and Applications, Cambridge, Cambridge. This is detailed undergraduate book on engineering thermodynamics.
113. van Ness, H. C., (1983). Understanding Thermodynamics, Dover, New York. This is a short readable monograph from a chemical engineering perspective. It first appeared in 1969.

114. Vincenti, W. G., & Kruger, C. H., (1976). Introduction to Physical Gas Dynamics, Krieger, Malabar, Florida. This graduate text on high speed non-equilibrium flows contains a good description of the interplay of classical and statistical mechanics. There is an emphasis on aerospace science and fundamental engineering applications. It first appeared in 1965.
115. Lewis, G. N., & Randall, M., (1961). Thermodynamics and the Free Energy of Chemical Substances, Second Edition, McGraw-Hill, New York. This book, first published in 1923, was for many years a standard reference text of physical chemistry.
116. Liepmann, H. W., & Roshko, A., (2002). Elements of Gasdynamics, Dover, New York. This is an influential text in compressible aerodynamics that is appropriate for seniors or beginning graduate students. First published in 1957, it has a strong treatment of the physics and thermodynamics of compressible flow along with elegant and efficient text. Its treatment of both experiment and the underlying theory is outstanding, and in many ways is representative of the approach to engineering sciences fostered at the California Institute of Technology, the authors' home institution.
117. Maxwell, J. C., (2001). Theory of Heat, Dover, New York. This is a short readable book by the nineteenth century master, first published in 1871. Here, the mathematics is minimized in favor of more words of explanation.
118. Moran, M. J., Shapiro, H. N., Boettner, D. D., & Bailey, M. B., (2014). Fundamentals of Engineering Thermodynamics, Eighth Edition, John Wiley, New York. This is a standard undergraduate engineering thermodynamics text, and one of the more popular. First published in 1988, it has much to recommend it including good example problems, attention to detail, good graphics, and a level of rigor appropriate for good undergraduate students.
119. Morse, P. M., (1969). Thermal Physics, Second Edition, Benjamin, New York. This is a good undergraduate book on thermodynamics from a physics perspective. It covers classical theory well in its first sections, then goes on to treat kinetic theory and statistical mechanics. It first appeared in 1964.
120. Müller, & Ruggeri, T., (1998). Rational Extended Thermodynamics, Springer-Verlag, New York. This modern, erudite monograph gives a rigorous treatment of some of the key issues at the frontier of modern

continuum thermodynamics.

121. Müller & Weiss, W., (2005). Entropy and Energy, Springer-Verlag, New York. This is a unique treatise on fundamental concepts in thermodynamics. The author provide mathematical rigor, historical perspective, and examples from a diverse set of scientific fields.
122. Müller, (2007). A History of Thermodynamics: the Doctrine of Energy and Entropy, Springer-Verlag, Berlin. The author gives a readable text at an advanced undergraduate level which highlights some of the many controversies of thermodynamics, both ancient and modern.
123. Müller, & Müller, W. H., (2009). Fundamentals of Thermodynamics and Applications: with Historical Annotations and Many Citations from Avogadro to Zermelo, Springer, Berlin. The author presents an eclectic view of classical thermodynamics with much discussion of its history. The text is aimed at a curious undergraduate who is unsatisfied with industrial-strength yet narrow and intellectually vapid textbooks.
124. Nernst, W., (1969). The New Heat Theorem: Its Foundation in Theory and Experiment, Dover, New York. This monograph gives the author's exposition of the development of the third law of thermodynamics. It first appeared in English translation in 1917 and was originally published in German.
125. Paolucci, S., (2016). Continuum Mechanics and Thermodynamics of Matter, Cambridge, New York. This graduate level monograph has an extensive discussion of how thermodynamics fits within the broader structure of continuum mechanics.
126. Schrodinger, E., (1989). Statistical Thermodynamics, Dover, New York. This is a short monograph written by the one of the pioneers of quantum physics, the co-winner of the 1933 Nobel Prize in Physics. It is based on a set of lectures delivered to the Dublin Institute for Advanced Studies in 1944, and was first published in 1946.
127. Shapiro, H., (1953). The Dynamics and Thermodynamics of Compressible Fluid Flow, Vols. I and II, John Wiley, New York. This classic two volume set has a comprehensive treatment of the subject of its title. It has numerous worked example problems, and is written from a careful engineer's perspective.
128. Smith, M., van Ness, H. C., & Abbott, M., (2004). Introduction to Chemical Engineering Thermodynamics, Seventh Edition, McGraw-Hill, New York. This is probably the most common undergraduate text in thermodynamics in chemical engineering. It first appeared in 1959.

It is rigorous and has went through many revisions.

129. Sommerfeld., (1956). Thermodynamics and Statistical Mechanics, Lectures on Theoretical Physics, Vol. V, Academic Press, New York. This is a compilation of the author's lecture notes on this subject. The book reflects the author's stature of a leader of theoretical physics of the twentieth century who trained a generation of students (e.g., Nobel laureates Heisenberg, Pauli, Debye, & Bethe). The book gives a fine description of classical thermodynamics with a seamless transition into quantum and statistical mechanics.
130. Tester, W., & Modell, M., (1997). Thermodynamics and Its Applications, Third Edition, Prentice Hall, Upper Saddle River, New Jersey. First appearing in 1974, this entry level graduate text in thermodynamics is written from a chemical engineer's perspective. It has a strong mathematical development for both classical and statistical thermodynamics.
131. Truesdell, A., (1980). The Tragicomic History of Thermodynamics, 1822–1854, Springer Verlag, New York. This idiosyncratic monograph has a lucid description of the history of nineteenth century thermal science. It is written in an erudite fashion, and the reader who is willing to dive into a difficult subject will be rewarded for diligence by gain of many new insights.
132. Truesdell, A., (1984). Rational Thermodynamics, Second Edition, Springer-Verlag, New York. This is an update on the evolution of classical thermodynamics in the twentieth century. The book itself first appeared in 1969. The second edition includes additional contributions by some contemporaneous leaders of the field.
133. Turns, S. R., (2000). An Introduction to Combustion, Second Edition, McGraw-Hill, Boston. This is a senior-level undergraduate text on combustion which uses many notions from thermodynamics of mixtures. It first appeared in 1996.
134. Turns, S. R., (2006). Thermodynamics: Concepts and Applications, Cambridge, Cambridge. This is detailed undergraduate book on engineering thermodynamics.
135. van Ness, H. C., (1983). Understanding Thermodynamics, Dover, New York. This is a short readable monograph from a chemical engineering perspective. It first appeared in 1969.
136. Vincenti, W. G., & Kruger, C. H., (1976). Introduction to Physical Gas Dynamics, Krieger, Malabar, Florida. This graduate text on high speed

non-equilibrium flows contains a good description of the interplay of classical and statistical mechanics. There is an emphasis on aerospace science and fundamental engineering applications. It first appeared in 1965.

137. Yoshida, S., Nonaka, I., Aoki, A., & Nishino, (1996). *J. Power Sources, 60*, 213.
138. Kibi, Y., Saito, T., Kurata, M., Tabuchi, J., & Ochi, A., (1996). *J. Power Sources, 60*, 219.
139. Diederich, P., Milani, A., Schneuwly, R., & Gallay., (1999). *Applied Physics Letters 75,* Nr 17, 2662.
140. Tanahashi, A., Yoshida, A., Nishino., (1990). *J. Electrochem. Soc. 137*, 3052.
141. Mayer, S. T., Pekala, R. W., & Kaschmitter, J. L., (1993). *J. Electrochem. Soc. 140,* 446.
142. Iijima, S., (1991). *Nature 354,* 56 [10] Ebbesen, T. W., (1992). *Nature 358,* 220 de Levie, R., (1963). "On porous electrodes in electrolyte solutions: I. Capacitance effects." *Electrochimica Acta 8*(10), 751–780.
143. de Levie, R., (1964)., "On Porous Electrodes in Electrolyte Solutions: IV." *Electrochimica Acta 9*(9), 1231–1245.
144. Posey, F. A., & Morozumi, T., (1966). "Theory of Potentiostatic and Galvanostatic Charging of the Double layer in Porous Electrodes." *J. of the Electrochemical Society 113*(2), 176–184.
145. Johnson, A. M., & Newman J., (1971). "Desalting by Means of Porous Carbon Electrodes." *J. of the Electrochemical Society 118*(3), 510–517.
146. Fritts, D. H., (1997). "An analysis of Electrochemical Capacitors." *J. of the Electrochemical Society 144*(6), 2233–2241.
147. Lin, C., Ritter, J. A., et al., (1999). "Correlation of Double-layer Capacitance with the Pore Structure of sol-gel Derived Carbon Xerogels." *J. of the Electrochemical Society 146*(10), 3639–3643.
148. Endo, M., Maeda, T., et al., (2001). "Capacitance and Pore-size Distribution in Aqueous and Nonaqueous Electrolytes using Various Activated Carbon Electrodes." *J. of the Electrochemical Society 148*(8), A910–A914.
149. Vix-Guterl, C., Frackowiak, E., et al., (2005). "Electrochemical Energy Storage in Ordered Porous Carbon Materials." *Carbon 43*(6), 1293–1302.

150. Celzard, A., Collas, F., et al., (2002). "Porous Electrodes-based Double-layer Super Capacitors: Pore Structure Versus Series Resistance." *J. of Power Sources 108*(12), 153–162.
151. Kim, H., & Popov, B. N., (2003). "A Mathematical Model of Oxide/carbon Composite Electrode for Super Capacitors." *J. of the Electrochemical Society 150*(9), A1153A1160.
152. Zheng, J. P., (2003). "The Limitations of Energy Density of Battery/double-layer Capacitor Asymmetric Cells." *J. of the Electrochemical Society 150*(4), A484–A492.
153. Zheng, J. P. (2005). "Theoretical Energy Density for Electrochemical Capacitors with Intercalation Electrodes." *J. of the Electrochemical Society 152*(9), A1864–A1869.
154. Zheng, J. P., J. Huang, et al. (1997). "The Limitations of Energy Density for Electrochemical Capacitors." *Journal of the Electrochemical Society 144*(6), 2026–2031.
155. Pell, W. G., & Conway, B. E., (2001). "Analysis of Power Limitations at Porous Super Capacitor Electrodes under Cyclic Voltammetry Modulation and dc charge." *J. of Power Sources 96*(1), 57–67.
156. Conway, B. E., & Pell, W. G., (2002). "Power Limitations of Super Capacitor and Capacitance Distribution Operation Associated with Resistance in Porous Electrode Devices." *J of Power Sources 105*(2), 169–181.
157. Kim, I. H., Kim, J. H., et al., (2005). "Electrochemical Characterization of Electrochemically Prepared Ruthenium Oxide/carbon Nanotube Electrode for Super Capacitor Application." *Electrochemical and Solid State Letters 8*(7), A369–A372. Amatucci, G. G. (2006). Private communication.
158. Xing, J., Vatamanu, O., Borodin, D., & Bedrov., (2013). On the Atomistic Nature of Capacitance Enhancement Generated by Ionic Liquid Electrolyte Confined in Subnanometer Pores. *J. Phys. Chem. Lett. 4,* 132–140.
159. Marsh, H., & Rodriguez-Reinoso, F., (2006). Activated Carbon, Elsevier Ltd., Oxford, pp.243–365.
160. Lillo-Ródenas, M. A., Lozano-Castelló, D., & Cazorla-Amorós, D., (2001). A Linares-Solano. Preparation of Activated Carbons from Spanish Anthracite II. Activation by NaOH. *Carbon 39,* 751–759.

161. Maciá-Agulló, A., Moore, B. C., Cazorla-Amorós, D., & Linares-Solano. A., (2004). Activation of Coal tar Pitch Carbon Fibers: Physical Activation vs. Chemical Activation. *Carbon 42,* 1367–1370.

162. P. L. Walker Jr., Taylor, R. L., & Ranish, J. M., (1991). An Update on the Carbon-oxygen Reaction. *Carbon 29,* 411–421.

163. Quinn, F., & Holland, J. A., (1991). Carbonaceous Material with High Micropore and Low Macropore Volume and Process for Producing Same, US patent 5071820.

164. Py, X., Guillot, A., & Cagnon, B., (2003). Activated Carbon porosity Tailoring by Cyclic Sorption/decomposition of Molecular Oxygen. *Carbon 41,* 1533–1543.

165. Conesa, A., Sakurai, M., & Jr Antal, M. J., (2000). Synthesis of a High-yield Activated Carbon by Oxygen Gasification of Macadamia Nut shell Charcoal in hot, Liquid Water. *Carbon 38,* 839–848.

166. Tam, S., Jr Antal, M. J., Jakab, E., & Varhegyi, G., (2001).Activated Carbon from Macadamia Nut Shell by Air Oxidation in Boiling Water. *Ind. Eng. Chem. Res. 40* 578–588.

167. Benson, B. B., Krause, D., & Peterson, M. A., (1979). The Solubility and Isotopic Fractionation of Gases in Diluted Aqueous Solution. I. Oxygen. *J. Solution Chem. 8* 655–690.

168. Ue, A., Murakami, & Nakamura, S. J., (2002). A Convenient Method to Estimate Ion Size for Electrolyte Materials Design. *J. Electrochem. Soc. 149* A1385–A1388.

169. Ania, C. O., Pernak, J., Stefaniak, F. Raymundo-Piñero, E., & Béguin, F., (2009). Polarization Induced Distortion of Ions in the Pores of Carbon Electrodes for Electrochemical Capacitors. *Carbon 47,* 3158–3166.

170. Aurbach, Levi, M. D., Salitra, G., Levy, N., Pollak, E., & Muthu, J., (2008). Cation Trapping in Highly Porous Carbon Electrodes for EDLC cells. *J. Electrochem. Soc. 155* A745–A753.

171. Chen, Y., Zhang, X., Zhang, D., & Ma, Y., (2012). High Power Density of Graphene-based Super Capacitors in Ionic Liquid Electrolytes. *Materials Letters, 68*, 475–477.

172. Tsai, W. Y., Lin, R., Murali, S., Zhang, L. L., McDonough, J. K., Ruoff, R. S., Taberna, P. L., Gogotsid, Y., & Simon, P., (2013). Outstanding Performance of Activated Graphene based Super Capacitors in Ionic Liquid Electrolyte from-50 to 80°C. *Nano Energy, 2*, 403–411.

173. Wei, D., & Ng, T. W., (2009). Application of Novel Room Temperature Ionic Liquids in Flexible Super Capacitors. *Electrochemistry Communications, 11*, 1996–1999.

174. Chatterjee, S., Rachel Carter, R., Oakes, L., Erwin, W. R., Bardhan, R., & Pint, C. L., (2014). Electrochemical and Corrosion Stability of Nanostructured Silicon by Graphene Coatings: Toward High Power Porous Silicon Super capacitors. *J. Phys. Chem. C, 118* (20), 10893–10902.

175. Kim, J., & Kim, S., (2014). Preparation and Electrochemical Property of Ionic Liquid-Attached Graphene Nanosheets for an Application of Super Capacitor Electrode. *Electrochimica Acta. 119*, 11–15.

176. Maoa, L., Li, Y., Chi, C., Chan, H., & Wu, J., (2014). Conjugated Polyfluorene Imidazolium Ionic Liquids Intercalated Reduced Graphene Oxide for High Performance Super Capacitor Electrodes. *Nano Energy, 6*, 119–128.

177. Trigueiro, J. P. C., Lavall, R. L., & Silva, G. G., (2014). Super Capacitors based on Modified Graphene Electrodes with Poly(ionic liquid). *J. of Power Sources, 256*, 264–273.

178. Kim, T. Y., Lee, H. W., Stoller, M., Dreyer, D. R., Bielawski, C. W., Ruoff, R. S, & Suh, K. S., (2011). High Performance Super Capacitors Based on Poly(ionic liquid)-Modified Graphene Electrodes. *ACS Nano, 5*(1), 436–442.

179. Liu, W., Ju, C., Jiang, D., Xu, L., Mao, H., & Wang, K., (2014). Ionic Liquid-Assisted Grown of beta-nickel Hydroxide Nanowires on Reduced Graphene oxide for High-Performance Super Capacitors. *Electrochimica Acta.,143*, 135–142.

180. Kim, J., & Kim, S., (2014). Surface-Modified Reduced Graphene Oxide Electrodes for Capacitors by Ionic Liquids and Their Electrochemical Properties. *Applied Surface Science. 295*, 31–37.

181. Yang, X., Zhang, L., Zhang, F., Zhang, T., Huang, Y., & Chen, Y., (2014). A high-Performance all-Solid-state Super Capacitor with Graphene-doped Carbon Material Electrodes and a Graphene Oxide-doped Ion Gel Electrolyte. *Carbon. 72*,381–386.

182. Tamilarasan, P., Ramaprabhu, S., (2013). Graphene Based all-Solid-State Super Capacitors with Ionic Liquid Incorporated Polyacrylonitrile Electrolyte. *Energy, 51*, 374–381.

183. Pandey, G. P., Rastogi, A. C., Westgate, C. R., (2014). All-Solid-state Super Capacitors with Poly(3,4-ethylenedioxythiophene)-Coated Carbon Fiber Paper Electrodes and Ionic Liquid Gel Polymer Electrolyte. *J. of Power Sources, 245*, 857–865.

INDEX

A

Accelerating Rate Calorimeter (ARC) 28
Accessible optional battery frameworks 19
American National Standards Institute (ANSI) 174
Antacid power modules (AFC) 91
Approach demonstrate 53
Assistant power units (APUs) 36
Astounding gravimetric vitality thickness 10
Asymmetric hybrids 205

B

Battery achievement 70
Boric corrosive arrangements 160
Business intrigue 10

C

Calorimeter temperature 29
Carbon-based materials 39
Carbon Capture and Storage (CCS) 148
Carbon dioxide-prompted electrolyte 130
Cathode materials 40, 42
Cell temperature 132
Ceres Power Limited 125
Charge-release process 59, 65
Charge-stockpiling gadget 180
Civil Aviation Authority (CAA) 72
Composite cathodes coordinate carbon-based materials 204
Consumption dependability 133
Contain methacrylate 144
Control electric vehicles 26
Coordinate carbon power device 78
Coordinate methanol power device 78
Couple potential developments 71

D

Deliberate warmth transition information 32
Demonstrating approach 46
Diminished cycling dependability 201
Disentangles warm administration 124
Dynamic material 43
Dynamic molecule 48, 50, 51, 52, 53

E

Earthenware capacitor 162
Electrical conductivity 206
Electrical potential 53, 54, 55
Electrical twofold layer (EDL) 163
Electrical vitality 93
Electric/electrochemical twofold layer capacitor (EDLC) 180
Electric field particular electric potential 49
Electric twofold layer capacitor (EDLC) 165
Electric twofold layer (EDL) 177
Electric vehicle application 42
Electric vehicles (EV) 20
Electrochemical and thermodynamic fundamentals 80
Electrochemical capacitors (ECs) 163, 165
Electrochemical change procedures 85
Electrochemical changes of dynamic mass 166
Electrochemical half-cell responses 90
Electrochemically produce power 124
Electrochemical redox surface procedures 226
Electrolyte arrangement 182, 183
Electrolyte hypothesis 54, 55
Electrolyte store plate (ERP) 133
Electrolytic aluminum capacitors 160
Electrometric information 30
Electromotive power (EMF) 72
Electronic application 19
Electronic business advance 22
Electronic gadgets 10, 12
Electro synthetic 59
Enactment misfortune 93
Energy component execution 90
Energy components (FC) 78
Energy thickness 177, 194, 198
Entire arrangement of transport 55
Entropy change 118
Environmental change 163
Essential battery 10, 11, 12, 17
Essential battery frameworks 10
Essential electrochemistry standards 80
Essential power framework 72
Essential thermodynamic 80
Extraordinary premium 153

F

Faradaic capacitive execution 177
Federal Aviation Regulations (FAR) 72
Fixed lead-corrosive 5
Focus subordinate ionic conductivity 48
Functional energy 121
Fundamental thermodynamic 90

G

Gem structure of lithium-intercalated graphite 63
Gibbs vitality information 58
Gluing dynamic material 37
Green Electricity (GEL) 214

H

Half electric vehicle (HEV) 65
Harsh dependable guideline 108
High temperature energy 110
High-temperature energy components 93
Huge scale vitality stockpiling industry 78
Hybrid capacitors endeavor 203
Hydrogen oxidation response (HOR) 139
Hydrogen-oxygen 135
Hypothetical potential 58, 59
Hypothetical thermodynamic cycle 113

I

Idle Stop Start (ISS) 225
Industrial facility 85
Innovation of lithium intercalated materials 21
Intelligent Energy (IE) 154

J

Joint Airworthiness Authority (JAA) 72

L

Laboratory of Electromagnetic and Electronic Systems (LEES) 215
Lanthanum manganite 126
Leading polymer positive terminal 204
Leading polymers 201
Likewise accommodate the correct extremity 14
Liquid carbonate electrolyte 144
Liquid carbonate energy component 78
Liquid carbonate power modules 92, 119
Lithium energy 41
Lithium-particle battery 71
Lithium-particle polymer battery 19
Lithium polymer 19
Lithium-polymer electrochemistry 20
Little fixed rechargeable battery 18
Lower framework productivity 98

M

Memory reinforcement 10, 11, 12, 15
Metal-air batteries 15
Moderately negative potential 7
Multi-walled carbon nanotubes (MWNTs) 206

N

Net calorific esteem (NCV) 194, 197
Nickel-Metal Hydride 19
Nippon Electric Company (NEC) 163
Non-development work yield 58
Non-equilibrium thermodynamics system 47
Nonlinear comparable circuit dis-

play 65
Nonlinear frameworks 79
Notwithstanding regular electrostatic 190
Numerous application 11

O

Open circuit voltage (OCV) 117

P

Particular framework 98
Particular orbital-warm programming 65
Peltier coefficient 53
Perfect proficiency 112
Permeable microstructure 126
Petroleum derivative 4
Phone potential misfortune 122
Phosphoric corrosive 78, 81, 88, 129, 132, 134
Phosphoric corrosive power devices (PAFC) 91
Photovoltaic power yields 84
Pinnacle Research Institute (PRI) 163
Plentiful non-platinum-assemble metals 137
Polyanion-based cathode applicants 40
Polymer electrolyte fuel cells (PEMFC) 91
Polymer Electrolyte Membrane Fuel Cell (PEMFC) 135
Polymer Electrolyte Membrane (PEM) 135
Poly pyrrole framework 207
Poly pyrrole polymer based anode 204
Poly vinyl liquor 144
Potential current reaction 93
Power device electrolyze frameworks 152
Power device process 112
Power module innovation 83
Power module working parameter 98
Power plant operation 98
Proportionate arrangement protection 170
Proton trade film energy component 78
Proton trade layer (PEM) 155
Pseudo-faradic response originates 202
Punctual rechargeable 38

Q

Qualify fixed nickel-cadmium 5
Quaternary ammonium (QA) 138
Quickened rate calorimeter 64

R

Reconnaissance hardware 12
Rectangular cyclic voltammetry 183
Redox flow battery (RFB 78
Regarding vitality thickness 197
Reversible cell potential 118, 119, 120, 121, 122
Reversible cell proficiency 119
Reversible vitality transformation framework 82
Ruthenium dioxide 164

S

Self-release of a super capacitor 217
Simultaneous advancement of elec-

tronic innovation 11
Single walled carbon nanotubes (SWNTs) 206
Solid Oxide Fuel Cells (SOFC) 153
Solid Polymer Electrolyte (SPE 20
Soluble battery 12, 14
Soluble manganese dioxide 12, 13
Standard hydrogen terminal (SHE) 26
Stationary energy 152
Straightforward 165
Strong electrolyte interface (SEI) 55
Strong oxide energy components (SOFC) 92
Strong oxide fuel cell 78
Strong oxide power devices (SOF-Cs) 124
Substantial scale applications 226
Super capacitors accompany 218
Superficial current thickness 62
Surface oxides 202
Synthetic vitality 36, 58, 66

T

Temperature appropriation 46
Temperature-modified desorption (TPD) 202
Temperature versus profundity of-release 65
Thermodynamic 28, 29, 33
Thermodynamic execution 82, 83
Thermodynamic restriction 17
Traditional dielectric 196

U

Uninterruptable Power Supplies (UPS) 221
Unmanned aeronautical vehicles (UAV) 155
Utilization energy 155
Utilize sulfuric corrosive 160
Utilizing nickel hydroxide 205

V

Vented lead-corrosive (VLA) 4
Vented nickel-cadmium (VNC) 4
Vitality misfortune 118, 120
Vitality stockpiling framework 17, 18, 21, 22
Vitality thick electrochemical cells 16
Vitality thickness 70
Vitality transformation gadget 112
Volume quickening rate calorimeter 66

W

Water administration 78
Water-gas move response (WGS) 147
Working temperature 104, 106
Worldwide temperature alteration 80

Y

Yttria doped zirconia (YSZ) 125